A Long White Scarf

Flying with Ripley Miller

by Julia Doten

A Long White Scarf
Flying with Ripley Miller

by Julia Doten

Edited by Robin Stratton

©2005 by Julia E. M. Doten
All rights reserved. No part of this book may be reproduced in whole or in part in any form without permission from the author.

ISBN: 0-9753211-2-9
Library of Congress Control Number: 2005902893

Printed in the United States of America
by King Printing, Lowell, MA

Published by

Branch Books *stories from the family tree* **WILMINGTON, MA**

www.home.earthlink.net/~branchbooks/

In cooperation with

Big Table Publishing Company, Newton, MA
www.BigTablePublishing.com

Cover Photograph by Kenneth J. Miller
Cover Design by Erica Doten and Julia Doten

To my family: "I thank my God upon every rememberance of you." Phil. 1:3

Thanks, Dad, for all the great photographs.

Thank you, Anna, for lending your reading, writing, and design skills at a moments notice.

ACKNOWLEDGMENTS

Many special thanks to:

Tracie Swiecki, who started this.

Jessica Lavey, for invaluable memories.

Billie Downing, for her technical help and always being able to find just what I needed.

Jean Batchelder, for her constant encouragement, adept proofreading, speedy return letters and positive attitude in all things.

Dad, Janice, Jeff, Caroline, Gramma, Warren, Jessica, Billie, Jean and Lillian for their thoughtful insight and recollections.

Friends and relatives who shared their memories of the past.

Robin Stratton, for her conscientious editing, honest criticism and praise.

Rebecca Beraldi, for her proofreading skills.

...for Ripley's family and friends.

PROLOGUE

Taking off from Tew-Mac

Lately, faded memories of my mother and her flying have been coming back to me, sketchy at first and dragged to the surface, then fuller and rising unbidden. Those memories were tucked away when childhood was left behind, and replaced with jobs, car repairs, a husband, a family. But now, stopping on the side of Route 38 where Tew-Mac Airport once bustled with activity, I stare at a row of newly-built condominiums -- and with very little effort I can imagine them gone and see again the blacktop runway.

On the left and right of the runway are parallel taxi ways, their paved surface showing bumps, cracks and weeds befitting a less important use. Planes neatly lined up along the outside of each taxi way, noses pointed in, form a gauntlet that encloses the blacktop and scruffy grass strips in between. To my immediate left a row of people lean against a chain link fence, eating ice cream and watching the planes come and go; one at the far end of the taxi way, another 1,000 feet high in the landing pattern, and yet another waiting its turn in a holding position before takeoff, cycling the engine. A whoosh of air ruffles my hair as a plane glides overhead to touch down for a landing.

There are still some things I want to do before Mom returns from giving a flying lesson. Skipping along the dividing fence that separates the end of the runway from Route 38, a thin line of patchy weeds pretending to be grass at my feet, I joyfully head toward the diner. Mom doled out some spare change earlier in the day, and I've held on to it until boredom, hunger, and suspense drives me to spend it on a soft serve vanilla ice cream.

Now with fingers sticky from the dripping mess, I traipse toward the one-room, one-story airport terminal where a little lavatory next to the counter separates workspace from customer space. Washing my hands with a drop or two of water and leaving stickiness on all the door handles I touch, I casually trail my hand along the countertop as I make my way back outside.

I stick my worn Converse® sneakers with the rubber toes into the holes in the chain link fence, and walk my way back toward the diner, holding precariously to the flexing top. Jackpot! I jump down to pick up a Bazooka® bubble gum comic strip lying on the ground, that I read and put in my pocket. Imagining I am **Harriet the Spy** *in the book by Louise Fitzhugh, I sit on the edge of a red painted picnic table and listen in on the conversations of strangers.*

As a little latchkey girl at the airport, I thought I knew more about flying than most. The independence I had while waiting for my mother helped to give me that attitude. However, most of my time was spent not learning about airplanes and flying, but waiting.

PROLOGUE

Waiting for Saturdays and summer days, when the airport owner's daughter would come and play with me. Waiting for my mother to part with some change so I could have a special, sugary treat of soda, gum or ice cream. Waiting for her to finish working, so we could go home. Waiting that was interminable as a child.

As an adult I can see how quickly the time flew by. My memories can conjure up the images of what this place once looked like, but cannot hold those images for long. An American flag snapping to attention brings me back to the present. It has replaced the orange windsock which hung high atop a telephone pole, and sod has been laid over the dry dirt that I used to draw in with my fingers and accidentally smudge across my face. Not far away from the glaring sign for the golfing community -- in the McDonalds parking lot -- a car alarm is alternately beeping, then honking.

With change there comes gradual forgetfulness. Who will remember the dreams my mother forged here and brought to fulfillment? There is much I never knew about her, and more than I can recall, so I am compelled to ask -- who was this woman I call Mom?

ONE

The front page of the two cent *Wakefield Daily Item* brought the more immediate news to the townspeople. A proposed ninety cent tax increase received only a slightly larger headline than the story of the special chicken dinner being planned by the new management of Howard Johnsons at Lake Quannapowitt for Saturday, March 13, 1937. Betty and George Conner had planned to go to the dinner; but as it turned out, their second child, Elizabeth, chose that day to come into the world.

CHAPTER ONE

The family lived in the Greenwood section of Wakefield, Massachusetts with daughters Caroline, Elizabeth, and a few years later, Sara. Betty was a kindergarten teacher before having children, and then became a weekend waitress; and George was an optometrist. Not wanting two Betties in the same household, they called Elizabeth by her middle name Ripley from infancy; the Ripley coming from her great grandmother's maiden name.

Ripley was an adventurous toddler, causing her mother to tie a piece of rope around her waist and attach the other end to a group of low trees growing near the house. She had shade and a little place of her own to play, but still she spent most of her time working that knot. On one occasion she did manage to untie the rope and wandered off. Betty, in a panic, called her name and searched the yard and woods. She at last found Ripley down off their hillside and across the street.

Their home on High Street was at the end of a precipitous driveway carved into rock and earth, that made several S turns before leveling out at the top. The woods that covered the landscape completely hid the bungalow with the fieldstone foundation, and the foundation of the garage that remained long after the building was knocked down by a hurricane.

When Ripley began first grade the teacher called her Elizabeth and though a little confused, Ripley insisted, "My name is *Ripley*." She was a tomboy as a child, participating in kickball, roller skating, biking, and general rough and tumble about the yard. Even so, the uphill walk home from school was arduous, and Rip thought she had come up with a novel idea in the second grade by jumping onto the back of the garbage truck, holding on for dear life as it chugged its way up the hill.

Growing up during the second World War left a deep impression on her. Between the poor state of the family finances and the rationing of the War, they lived with scarcity and made do with what they had. The sisters considered themselves lucky to have been given some grain sacks printed with a pretty flower pattern which their mother made into skirts. Their car was up on blocks to protect the tires, and since gas and money was in such short supply there was no chance of using it anyway. Families were given an allotted portion of stamps with which to buy food, and Betty used her gardening skills to grow fruit and vegetables, canning and freezing for the coming winter. Ripley never forgot her mother stretching meals and money and clothes, and resolved that her life would be different.

Like her mother, she was outgoing and personable, creative and intelligent, independent, self-assured and doggedly determined. Though these traits became a benefit to her, during her teenage years the two similar personalities clashed as Ripley, yearning to sprout wings and fly free, continually defied her mother's authority.

As a sophomore at Wakefield High she signed her book "E. Ripley Conner ," and by the time she graduated the "E." was gone. She was active in band and orchestra, playing clarinet, and was on the school newspaper and yearbook staffs. In her senior yearbook, she wrote that she had a suppressed desire to parachute from an airplane. She dared to do things others wouldn't; like going to nightclubs underage, smoking cigars on the train into Boston, changing her name at will -- telling people (especially young men), she was Elizabeth, or Rip or Lee.

She had a calendar full of school activity reminders, party dates, her work schedule at the local diner, and who she had a date with that evening. On each page she wrote doodles such as, "Who do ya love?" Brian (later crossed out, as were all the others), Andy, Leonard, Bob, Don, "Senior yr and out!"

CHAPTER ONE

When?" with a drawing of a diamond ring, and "Hope" with a drawing of a chest of drawers, and "When?" with a drawing of a wedding invitation.

The pressure to meet a boy and marry, raise a family and subsequently live happily was an inescapable influence in the 1950s. Marriage meant a way out of the childhood home for a young woman. Wages for women were so low that most were unable to earn enough to support themselves. As for Ripley, she had more than just a desire for independence.

Her father had taken control of the business he worked for when the owner died. He found he was good at his job, but bad at managing the business. As the company began to fail, he began to drink; and spoiled whatever relationship he'd had with his wife and three daughters. By the time Ripley was in high school the marriage had deteriorated, becoming a matter of shared living space and obligations.

Ripley wanted to get away and start a life of her own. One that would not have the same problems her parents had.

During the summer after graduation Ripley studied at the Comptometer School in Boston, adding to her secretarial skills by mastering the comptometer, an early model of the adding machine. After completing the course of study, she worked full time as a tax clerk at Hale and Dorr in Boston, in addition to waitressing weekends at the diner. There were very few options for women at that time, Ripley knew. You were a wife and mother, a nurse, schoolteacher, secretary or waitress.

"It was her eyes… and her legs." In the raw cold of March 1956, Ken Miller stopped for a warming bowl of clam chowder, and found himself returning for the waitress with the sparkly brown eyes. He was the experienced, handsome, twenty-eight year old blue-eyed, blond customer, and she, barely nineteen, the dark-haired, classy-looking waitress. She showed her dislike of clam chowder and her attraction to him, by making faces as he ate. Without the appeal, she would have served the soup and distantly moved on to the next customer.

Ripley had enough experience and confidence by this time to scorn tradition, and ask Ken on their first date, to an Easter Service at the Cathedral of the Pines -- an outdoor church under towering pine trees in Rindge, New Hampshire.

Kenneth brought along his sister Joyce -- who was the same age as Ripley -- to set Betty and George at ease as he whisked their daughter away early on Sunday morning. The interest Ken and Ripley had toward each other grew stronger, and other dates followed rapidly.

She was at work in Boston only a month later when Western Union delivered a telegram to her that read, "EVERYTHING I WANT AND NEED. ALL MY LOVE KEN."

Two months later on June 30, 1956, Ripley and Kenneth were married. At first they lived at Kenneth's family home along with his parents, his brother Bill and his wife, his sister Joyce -- and within a short time -- her new husband, Al. They decided to move into a small apartment while Ken built a ranch house on a nearby lot in Burlington.

The training Ripley had received after high school would no longer be needed once a family came along. And within a month or two, that is just what happened. She had already left her waitress work, and now left the office job as well, content in her new role as a pregnant housewife. Kenneth finished their new house shortly before Ripley gave birth to Jeffrey the following April.

Childbirth had become a medical procedure and a combination of drugs was used which produced a temporary loss of memory called twilight sleep.

Women would feel pain, but had little or no recollection of the pain or the events of the birth. When they awoke to a ruddy infant being placed in their arms, they must have thought, *Are you sure this is mine?*

In Ripley's *Modern Family Health Guide*, published in 1959, a new method of childbirth is discussed called "natural childbirth." The author is quick to add, however, that this method is used only when the mother chooses it and her doctor agrees that it will not cause any harm. Ripley had heard plenty of childbirth stories from her older sister and her husband's sisters, who between the five of them had sixteen children when she started on her first. She insisted on a natural childbirth, and had no pain medication during delivery.

The day Jeff was born, Kenneth quit his job and went to work for himself, opening a civil engineering and land surveying business in a converted garage at home. Ripley probably wondered, *How the hell are we going to make ends meet now?*

TWO

For housewives in the fifties when money was short and household needs long, friendships helped them to get through the days. Ripley met Jessica Lavey one night at a grassroots "political" meeting held in a neighbor's home. Seated in a circle, listening to the speaker who was beginning to sound like a droning summer insect, Jessica looked across the room and spied a young, attractive woman and her handsome husband.

They found they both lived on Brown Avenue, but at opposite ends; and Jessica walked past Ripley's house to take her two children to the bus stop. Ripley invited Jessica to stop for coffee on the return trip.

CHAPTER TWO

It was Rip's way to hold the coffee cup in both hands, putting the cup to her cheek to feel the warmth while Jessica recounted her morning. Living near each other was only one of the things they had in common, and they soon got to know each other well.

They shared ideals of strong family, country and the principles of truth and trust, but most of all, they shared humor. Ripley could find the funny side of every situation, sometimes to the point of shocking Jessica with her bold statements and actions. Once when driving together, Jessica came to an intersection and asked Ripley if there were any cars coming. Ripley said "No," and she paused long enough for Jessica to begin to pull out then added, "only trucks and busses."

Together they made plans to go to the drive-in theater, but didn't admit to not having enough money to do so. Ripley collected and cashed in the glass deposit bottles, only to bump into Jessica on the way out doing the same thing. As their friendship deepened, they joined a bowling team, and then the League of Women Voters.

They quickly fell into a routine of relying on each other for childcare, listening to problems and offering advice, or borrowing a bit of something they'd run out of. When Ripley was pregnant with her second child, she craved Popsicles®, and bought one every day from the ice-cream truck. Since the truck started at Jessica's end of the street, Ripley phoned her, using a pencil by the eraser end to turn the dial on her black rotary phone. "Has he come by yet?" she demanded. She eventually bought Popsicles® by the box, exclaiming about her son, "Thank God Jeff isn't tall enough to reach them."

That fall Ripley gave birth to her second child, a daughter she named Janice; and she gratefully accepted Jessica's offer to babysit Jeff while she was in the hospital. In the months following the delivery, she worried over her weight and her shape. She was exercising before the fitness craze begun by Jack LaLaine. In addition to his exercise records, she used her household chores to keep trim.

"Does my tummy stick out too much?" she asked Jessica, and Jessica watched her put the basket of laundry on the floor, pick up one piece at a time to be folded -- keeping her stomach muscles contracted all the while.

Hanging diapers on the line, hands red from the cold, she wished out loud for a dryer to ease her workload. Ken hinted at Christmastime that he was getting her something white, and she told Jess "Oh Lord, you don't suppose he's going to get me a new clothesline as a joke, do you?"

When she did get a new dryer, her laundry work eased considerably. Clothing hung hot from the dryer helped to simplify her ironing. No-wrinkle cotton had yet to be invented, and everything from shirts, skirts, handkerchiefs and pillowcases had to be pressed, so most housewives had a huge pile of ironing. Phyllis Diller -- an up and coming young comedian -- joked that the clothes at the bottom of her pile of ironing were so old she had to bury them in the back yard. Ripley kidded Jessica by offering to loan her a shovel.

Like many newlyweds, Ripley's housekeeping underwent a transformation during the first several years of marriage. Ken never knew what he would come home to; the house wild with clutter of mail, toys, and clothes while Rip triumphantly showed him the single organized drawer or closet she had labored on all day. It was Dottie Graham who helped her along the way.

Dottie was the wife of a local builder who hired Ken to do his engineering work. The men introduced the two women, and they quickly became close friends. Dottie wore her blonde hair pulled back into a barrette, and had blue eyes accentuated by a heavy

CHAPTER TWO

streak of robin's egg blue eye shadow. Both Rip and Dottie shared the same zesty sense of humor, as well as problems and stories of raising children of the same age.

When Ripley bemoaned her inability to clean house, Dottie came over to show her a few tricks, beheld the cluttered state of affairs and said, "You're never going to be able to clean until you pick up all the clutter!" Dottie embarrassed her into action.

Ripley had been raised in a busy household that had more than its fair share of clutter; her mother was a crafter and antique collector. For many years she supplied her own summer business in Kennebunk, Maine, aptly named "Shop in the Shed and Barn," where she sold antiques and homegrown vegetables. There were stacks of this or piles of that and unfinished projects on the floor, tables and countertops.

Ripley as a teen was a sloppy roommate for her neat and orderly sister Caroline, who eventually drew a line in white chalk on their bedroom floor to designate sides.

As a grown woman with a family of her own, Rip found her mother's type of household didn't appeal to her. She wanted to be organized, meticulous and time conscious, she just hadn't learned how. Once she figured out how to remove clutter, the entire house became sparse and remained that way.

Rip introduced Dottie and Jessica to each other, and soon the three shared their daytime trials by phone and over coffee as well as an occasional evening out.

One night the ladies had plans, and Ripley couldn't wait to get out. The phone rang shortly before she was set to leave. Before she could say hello, she heard Jessica's voice coming from the receiver.

"I can't go. My dog is sick, and I have to give her medication every hour," Jess said dejectedly. "Just go without me."

Half an hour later, Jessica heard a car pull up in the driveway. Rip and Dottie had come to pick her up anyway -- along with the dog.

"Only for you Jess, would we do this," Ripley said as they huffed and puffed the overweight and languishing cocker spaniel into the back of the station wagon for a night at the drive-in. Together Ripley, Jess and Dottie made a companionable trio, and gave each other the laughter they needed.

The dog got better.

Ripley loved babies and couldn't wait for her third. But after Janice was born she experienced the heartache of two miscarriages; and the sadness for her was compounded as she was barely able to find the time to grieve for those little lost souls. She had two toddlers demanding her attention -- Jeffrey was two and a half years old, Janice only one -- and Ken was busy as he built his business.

Ripley's older sister Caroline by this time had five children of her own and lived several towns away in Melrose. Her younger sister was just eighteen and graduating from Wakefield High School. She talked with Kenneth's sister Joyce often by phone, who had two toddlers as well and lived nearby. Jessica and Dottie offered what help they could, and she realized how blessed she was to have these strong friendships.

Ripley's next pregnancy was a healthy one, and when she made it past the critical first three months, she excitedly shared the news.

When Janice was two years old, I was born. The joy of a new baby, a family that felt complete, and her friendships that proved to be lifelong, sustained her through the next few years.

THREE

"Listen to this," Mom says to me as she holds her newspaper in front of her. "Fly-Away Feat by Women Pilots. Hundreds of letters bearing the new airmail postage stamp honoring Amelia Earhart were flown out from her birthplace of Atchison, Kansas on July 24, 1963. A letter reached forty-six of our capitals in two days, and over the next few days, reached many other places throughout the world. The 'fly-away' was accomplished with great fanfare by seven women pilots who were peers of Amelia's." I look at her blankly. "Who's Melia?"

Though over a month old, it's still news to this small-town paper in Burlington, Massachusetts. The article is strategically placed, surrounded on one side by the supermarket sale prices of the week and on the other, the weekly scores of the local bowling teams. Every woman in town is sure to see it.

I watch my mother fold the newspaper, set it down, pick up a washcloth and wipe up spots of spilled cereal and milk on the yellow formica kitchen table. The smell of burnt toast hangs in the air, and blackened crumbs litter the countertop. This is how Dad likes his toast, and he doesn't notice the mess he leaves behind as he kisses Mom's smooth cheek and saunters out the door to work. When he left today she had a smile on her face but beware her ferocious frown! Mom is five foot seven, and slender with long, gracefully curved legs. I think she is full breasted, judging by her softness when I need a hug, and also by the lacy bras I peek at hanging on the drying rack.

A crashing noise comes from my brother's room, followed by thumps and banging and the start of a wail. "They're at it again," Mom says under her breath, "and the day has barely begun." She steps over a stuffed toy abandoned on the floor and makes her way down the hall toward the mounting noise to settle a fight between Jeff, my six-year old brother and Janice, my five-year old sister.

I scurry to catch up and reach the bedroom doorway in time to see Mom pull Jeff off of Janice just as he's about to deliver a slap. Janice sticks her tongue out at him! I almost tell but Mom is busy propelling her out of his room and into the one we share; both beds already neatly made, with our pajamas folded inside cat shaped storage bags and set atop our pillows.

"Do as you were told, get dressed and stay out of your brother's room! You'll miss the bus again!" Mom says with a rising crescendo as she delivers a swat to Janice's fanny.

Back in the kitchen she finishes making lunches, brushes the crumbs from the countertop into the barrel and calls once more for Jeff and Janice.

"Let's go now!" Like tumbleweeds blown on the prairie, the four of us roll out the door in succession and make our way to the bus stop. Just in time. Mom sighs a breath of relief when the bus pulls away.

CHAPTER THREE

"Quiet at last," she smiles down at me,"just the two of us." We leisurely walk back home, allowing time to stop and push a pebble into the sticky tar on the road.

Once at home, Mom continues to pick up from breakfast as I color with crayons, but she is drawn back to her paper. Her eyes scan the picture of the female pilot standing next to her plane, smartly dressed, smiling, the wave of her hand frozen in time. Is this pilot ever surrounded by mundane household tasks? When her children are squabbling and fighting, or the bill collectors calling to demand a five-dollar payment? She compares herself to the unknown woman.

"Who am I?" she asks out loud. "You're my Mom," I answer, and follow her as she walks into her tiny bedroom. She lifts me up and drops me on the bed with a bounce, and I giggle. Closing the door, she looks into the mirror that hangs on the back side and as she peers at her reflection, I start to jump on the bed. "Look at me, Mommy, look, look, look," I say with each leap. But she doesn't look at me, she sees something else.

It's her first gray hair.

"Not one word!" Ripley said to Ken when he came home to find flour covering her, her apron and every surface of the kitchen. Piles of pie dough were thrown haphazardly in the wastebasket nearby, and bowls, measuring cups and spoons littered the countertop. He tiptoed out of the room, while she continued clattering away in the kitchen. The pie she served for dessert that evening received cautious praise from Ken. "This one is very good. I think you've got it!"

Ripley taught herself well and found she loved to bake, cheerfully making huge platters of different Christmas cookies to give to friends each year. Russian teacakes, round balls of crumbly delight covered with confectioner's sugar. Butter cookies squeezed out through a Sears and Roebuck cookie press in shapes of trees, stars, dogs and camels, all sprinkled with colored sugar. Round vanilla cookies with nutmeg on top, cocoa brown hermits loaded with raisins, and crispy oatmeal cookies, rolled into a log and wrapped in wax paper, that spent the night in the fridge before baking.

Ripley also baked delicious bread, using a large metal bowl that had a stirring arm with a round black knob to knead the dough. Manufactured by Landers, Frary & Clark of New Britain, Connecticut, the No. 4 Universal Bread Maker was awarded a gold medal at the 1904 St. Louis exposition. Ripley liked it because the effort of holding the pail crooked in one arm and stirring with the other kept her arm muscles in good shape. She made half a dozen loaves at a time, her willing hands punching the risen dough down, flour dusting her cheeks. Baking aromas filled the kitchen, the house, spilling out into the yard to where Kenneth's home office was located in the garage.

Not long after the bread came out of the oven, Kenneth and one of his clients came out of the office, and into the kitchen. "What's that we smell?" They knew perfectly well what it was -- and that a little flattery would get them a piece. Of course she'd tear off and butter hunks of steaming softness, chatting all the while. She enjoyed any chance to talk to adults after being with three children all day.

Once Ripley settled into a routine and learned a bit about cooking, she would make supper every day at the same time, when the show *Sea Hunt* starring Lloyd Bridges came on her black and white television. She admitted to Jessica, "I'm worried that I'll be at someone else's house and hear that cursed theme song and get up and start cooking dinner, like Pavlov's dogs' reaction."

With the help of her friends and the blessed passage of time, Ripley had reached a point where she was the epitome of the American housewife: attractive and well groomed, her children clean and mannerly, her home tastefully though meagerly decorated. She was an excellent cook, a loving, supportive wife, active in church and community affairs.

But was she happy with herself and her accomplishments?

The walls of pretense began crumbling down around her and countless other women like her who were pressured to fit into roles predetermined by American society. She worked hard to keep it all together, and loved the rewards of her family, but still something was missing. She wanted more.

When the time came for Ripley to grow and change, she was ready. It was October 8, 1963.

FOUR

    We live in a suburban town north of Boston, in a neighborhood where kids get together to play baseball in the empty lot next door in the summer, and go sledding at the hill down the road in the winter.
    Step through the side door in our ranch house into the kitchen, and you can see the hallway opening at the opposite end of the room leading to the dining room, the one bathroom, and three bedrooms.
    The living room is straight across the hall, and hosts a red brick fireplace, an overstuffed wing back chair and sofa, a Boston rocker, a large braided rug, and a black and white television with rabbit-ears antenna.

CHAPTER FOUR

Our television only receives three channels. We get to watch Davey and Goliath *on Sunday mornings, and sometimes* Gumby and Pokey *or* Jonny Quest. *There aren't a lot of shows for kids, and some Mom won't let us see, like* The Three Stooges. *She says we'd do those things to each other.*

Immediately to the left is the door leading to the cellar, down dark, narrow and steep stairs, the site of many of my nightmares. During the day some light filters through the high set windows, and renders the place barely cheerful. It is here that Mom stands me on a chair in front of the ironing board, showing me how to press pillowcases and handkerchiefs.

Rainy weather finds the three of us kids in the cellar, sometimes writing and performing plays, sometimes opening a gourmet restaurant. Mom is a regular customer. Squeezing into a child's chair in our make-believe restaurant, looking at the menu, she complains loudly over the exorbitant prices we charge. She pretends to eat our fake, plastic food, and pays with invisible money. Or attending a play, she tears her ticket in half, turning in one half and keeping the other, clapping and laughing at all the right places. She is always our best fan.

A hard day at play brings the daily evening whine from the three of us kids. "What's for supper? I'm hungry."

"Goobalie guts on toast," Mom replies. This always makes us stop and think. Softly we sing the song that children sing when they are together in the yard and think no one is listening. "Great big gobs of greasy, grimy gopher guts, mutilated monkey meat, little birdies dirty feet..."

In general, goobalie guts on toast means a mishmash of things in a casserole dish, slopped over bread or saltines; a regular is a can of tuna, a can of cream of mushroom soup, egg noodles and a can of peas. We pick out the mushrooms. And the peas.

In the fall, Mom makes apple pies by the dozen. Janice and I work with her, using a crayon to make tic marks in groups of five as a reminder of how many cups of apples she has peeled and cut. Once she mixes the sugar and cinnamon in we eat piece after piece until she crossly tells us to stop. I guess we can eat faster than she can peel! When she is done rolling the crust, she gives us the extra dough, and a little pie tart in comic imitation of her pie is created.

She also serves us her invention of saltines topped with cottage cheese and a dollop of grape jelly, for the times when our starving bellies can't wait for dinner. Always there are homemade cookies which make up for the days when she serves goobalie guts. And for breakfast, hot oatmeal with brown sugar melting on top.

Once a week milk is delivered to our kitchen door from Sunnyhurst Dairy in Stoneham. A wire basket holds six quart glass bottles with paper caps pressed across their tops. Jeff peels off one of the tops and plays with it, then pours a little milk across his oatmeal. Not too much! He doesn't want the brown sugar to get washed away.

Kenneth developed an interest in flying when his older brother Bobby became a pilot in World War II. As children during the Depression, a plane flying overhead was such a rare treat that everyone on their modest farm in Burlington stopped what they were doing to run out, dust kicking up from their bare feet, to watch and wave.

Now with his business firmly established and his family growing, he decided it was time to fulfill his dream. When Kenneth began taking flying lessons at Tew-Mac Airport in Tewksbury, Massachusetts, Ripley surreptitiously tagged along. Little did either know he would spark a dream for her as well. One day, remembering the excitement of a seaplane ride she and Ken had taken during their honeymoon, Ripley asked Ken if she could have an introductory lesson.

I'll humor her. I'm a modern, open-minded husband, Ken smirked to himself. After all, it was 1963!

CHAPTER FOUR

He was surprised when she landed and declared, "I'm going to get my pilot's license." Then she marched into the office to buy a logbook from the owner of the airport, Warren Hupper. Kenneth's openmouthed stare followed her. He waited a long time for her to return, almost too long. At last she emerged from the office, a black leather logbook in her hand, an exasperated look on her face. "Warren tried to talk me out of it!" she said.

Although Kenneth encouraged her to fly, she was the one who had to find money from the budget to be able to do so.

Complicating the task was Ken's inability to say no to salesmen. So Ripley had to deal with a once-a-month delivery of frozen meats and groceries, complete with a new freezer to keep them in, and five dollar payments for a lifetime or two. Or several dozen boxes of sickly green mints. Once he bought six designer watches from a door to door salesman, finding later they were missing the crown that allows them to be set and wound.

Many times he was short of cash for the company payroll, and had to "rob Peter to pay Paul," taking money from the household budget in order to keep employees paid, and sometimes going without pay for himself.

It was painful for Ripley to see her father's business practices mirrored on her husband. They argued over the business, over bills, over their inability to take a modest vacation. Seated at the table, a pile of bills on top, Ken asked Ripley to toss him a pencil. When she did he missed catching it, and the pencil point stuck into the bridge of his nose like an arrow.

Chagrined, she was all apologies; and unharmed, Ken started to laugh. She saw him that moment as she knew him to be, a man with a good sense of humor, slow to anger, generous to a fault -- and he didn't drink the way her father did. They laughingly agreed to keep shoveling sand against the tide.

Living on one income with a son in second grade, a daughter in first grade and me still at home, meant being a frugal manager of funds even without the flying lessons. The house was decorated plainly, with worn furniture and unadorned walls. Their bedroom was just as plain, not one picture on the wall, and barely enough room to walk between the double bed and the dresser.

There was no security in the availability of funds with a self-employed husband; "extras" were doled out as money was available. To that end, she cut the grocery bill by baking bread, the milkman's fee by adding water to the milk, and gratefully took all the veggies that her mother gave her. The sandwiches were spread a little thinner, the cuts of beef got a little tougher, and canned food was served a bit more often. Variations of goobalie guts on toast became a staple.

FIVE

Mom has great clothing style. A white knit suit with a pink shell, a red bubble body suit with a white wraparound skirt, tan stirrup pants with a blue sleeveless pullover and a Jackie Kennedy scarf and sunglasses. Holiday parties give her the opportunity to get together with friends and wear her dressiest clothes; sometimes a full-length turquoise skirt with a white ruffled blouse, or a short dark green velvet dress with a satin ribbon and rhinestone buckle that I love.

Dressing Dad is another matter. He doesn't give a hoot what he wears, and Mom oversees his wardrobe to be sure it is function appropriate, matching, and in style.

Somewhere along the way Dad lost his hair on the top, so at Mom's urging, he starts to wear a toupee. If I go into his room early enough, I can watch him cut a piece of double sided tape, stick it onto the bottom side of the toupee which is kept on a stand overnight, and then in one motion flip it up and over onto his head.

For special occasions Mom wears a partial wig called a fall, with ringlets of hair cascading down the back of her neck. I stand on her bed and help curl her hair with the curling iron; and holding my breath, watch in her dresser mirror as she clips on rhinestone diamond earrings. She stands back from me and smiles, and I exhale.

She began taking lessons in a Piper J-3 Cub with Warren, who was tall, fit, with gray-blue eyes and about the same age as Ripley. According to Warren, Ripley was hard on herself during training because she wanted to get it exactly right. She practiced landings over and over until Warren, exasperated, told her to let him out at the end of the runway!

For Ripley, flying was more than a hobby, it was a form of art. Her training in the job force and trials at housekeeping, child rearing, and cooking were all steps that led to this point in her life. If she could spend hours learning how to make a perfect pie crust, how much more could she accomplish now?

Her studies took increasingly more of her time and she saw her friends less and less.

"You're consumed by flying, " Jessica accused. "You don't even go to the movies anymore!"

Ripley tried to explain to her, "My eyes fly open at the crack of dawn and I rush to the window to see if the weather is flyable, a CAVU (clear air, visibility unlimited) day. I get my house work done as soon as possible so I can get airborne."

"But doesn't flying scare you, even a little?" asked Jessica.

Ripley shivered. "No, the only thing in life I am scared of is being on a ski slope; skiing scares me to death."

She told Jessica about the buildup to a new pilot's first solo flight, and the apprehension that accompanied the event."Everyone at the airport talks about their solo; if they bounced on their landing, or

CHAPTER FIVE

had to go-around again, and it makes me wonder, how will I do?" She continued confiding as the two sat companionably and watched us children play.

"When a student is about to solo, everyone stops what they are doing to watch, the instructor gives last minute reminders, and then the student is alone for a single take off and landing. There's not much time to dwell on fear."

But after Ripley soloed on January 17, 1964, with ten hours of dual instruction, she confessed to Jessica, "Knowing you are being watched, that you control a life or death situation, can be extremely intimidating. Your mind races with scenarios ranging from a perfect landing to a runway crash in a ball of fire; to flying around in the air frozen with fear until the plane runs out of fuel and you plummet to the ground." Jessica shook her head in disbelief while Ripley laughed.

"That takes place in the first ten seconds. Then your training takes over, you perform according to your best skills -- and lo and behold -- within a few short minutes you are safely back on the ground! Adrenaline causes a little aftermath shaking, but the thrill and congratulations are well worth it. You can't wait to do it again."

"Speaking of congratulations," Jessica said, and handed Ripley a small bag that she held unnoticed. "I looked everywhere for an aviator's helmet, but just couldn't find one," she explained, as Ripley opened the bag and pulled out a long white scarf.

A rite of passage was performed at Tew-Mac and many other airports. The student's shirttail was cut off after the first solo flight and labeled with their name, date and the instructors name using a black marker. It was then hung over the door frame at the office, beneath a giant wooden propeller lacquered to a shiny finish. But whatever Ripley wore the day of her solo flight fortunately remained on her back, a small concession to the fact that she was a woman. At any given time a dozen or so different colored and patterned pieces of cloth, about 12 inches square, were strung in a row like a line of drying laundry across the wall.

Rip was so enamored of flight that she enthusiastically supported the purchase a 1954 Cessna 170 that June. Kenneth had a plan to cut a hole in the floor and use the plane for aerial photography for his land surveying business.

"I can make money from the service and write off the expense of the plane," he reasoned. They both overlooked the fact that a camera, processing equipment and a darkroom were also needed; but Ken soon had those in hand as well, converting space in his office into a darkroom.

The plane, nicknamed a "taildragger" because the third wheel supporting the plane was located at the end of the tail, was a more difficult plane to fly than the Piper J-3. Unfortunately, Ripley had to start learning to fly the Cessna 170 almost from the beginning. The 170 had side by side seating and a wheel for turning in the air; the J-3 had tandem seating, front and back, and a stick for turning. But once again Ripley soloed after another ten hours of dual instruction.

Another dual lesson and she flew her second solo, then her third. "My instructor won't let me fly past Route 128, like a kid!" she sputtered one day. But soon she flew her first solo cross-country (a flight longer than 50 nautical miles from point of origin) to the omni (a radio navigation aid) at Gardener, Massachusetts and back.

She was taking lessons sometimes once a week, or three to five days apart, occasionally at a two-week interval. If it weren't for the financial restriction of paying for lessons, she would have trained nonstop!

CHAPTER FIVE

It was the beginning of a juggling and balancing act; now keeping the household running smoothly as well as fulfilling a longing for definition outside the traditional positions of a wife and mother.

The Cessna became the plane in which she earned her private pilot's license on April 18, 1965, after 74 hours of total flight time.

Ripley took her sister Caroline flying only once, for a trip along the New England coast. Caroline was terrified -- especially when they had to bank into a turn, and didn't want to go a second time. Ripley also invited Jess to go flying, but she declined.

"Machines malfunction. If they are on your desk, on the road, or in your laundry room, it's not too bad, just frustrating -- but in the *air*, you just plain go down," Jess said.

Rip reassured her. "Pilots are taught to be constantly on the lookout for open fields and safe places to put the plane down, just in case. With a small plane, enough altitude, and a well-trained pilot, there is a certain amount of glide that can be used to maneuver an emergency landing into a deserted field or parking lot."

Jessica wasn't convinced. And Ripley without realizing it sometimes compounded her fear. She elaborated on the trip she took to Las Vegas with Ken, his friend Bob Wignall, and Dottie Pink. Flying a fully loaded Beechcraft Debonair, the craft didn't have the ability to fly over the Rocky Mountains, so Bob piloted along the mountain range until he found a valley at a lower elevation to pass through. Ripley showed Jessica the pictures she took as they crossed.

Jessica commented, "Gee, a white plane in the snow. If anything had happened, no one would ever find you."

"We all planned to bleed a lot," was Ripley's quick reply.

SIX

A one-day trip to New York for the 1965 World's Fair, Peace Through Understanding, turns into three. We plan to land at a small airport called Flushing, but when we arrive, the airport is full and we are rerouted to La Guardia International. It is the difference between a tea cup and a bathtub. La Guardia is bigger than huge, our plane's crank radios won't stay on the channel Dad set them for, (they almost never do,) he has a splitting tension headache, and is such a wreck over the landing that once we shut down, he gets out of the plane and throws up.

CHAPTER SIX

At the fair, we walk and look at exhibits until we are exhausted, and at days end find we are unable to fly out because the area is surrounded by Big Ts -- thunderstorms. You can't fly through them, and they are so unpredictable that trying to get around them can box you in. We find a cheap motel and spent the night with little money left, and no toothbrushes or change of clothes.

The next day since the storms have not yet retreated completely, Mom and Dad decide to spend a little more time at the fair. Now totally broke and unable to pay for a second night at the motel, we must return home. Visibility remains poor, and a thick layer of clouds cover the setting sun. Dad decides there is enough ceiling to fly. Mom doesn't agree, but we start out anyway and cautiously make our way back north. We don't get very far.

As evening comes on, the sky is punctuated by lightning. Peering out the windows into ever-increasing darkness, we circle over and over, unable to find Connecticut's Waterville Airport. There is no choice but to find a suitable place to land, and fast. The dark blots out everything into formless shapes of varying degrees of black and gray. Dad spots a parking lot next to an apartment building and makes his way to the ground. His descent is too high, and he misses the parking lot, landing just beyond it in a grassy field too short to comfortably land in. He practically stands on the brakes to get the plane to stop at the edge of the woods beyond.

He claims he still has fingernail marks where Mom gripped his thigh.

The apartment building empties as those at home hear the sound of a plane engine; one woman at her kitchen sink exclaiming to her husband, "I just saw a plane fly by our window!" They come out to us with flashlights, concern and relief. The manager of the building lets us stay in an empty apartment where I learn the true meaning of "hardwood floor." He also shares a meager breakfast with us in the morning before we go outside to see where we landed, and determine our way back out.

God was truly with us that night. Unseen in the darkness the night before was a curb along the pavement. If we had landed in the parking lot and tried to cross into the field, we would have flipped over nose first. Two flattened wheel paths are visible in the grassy field. Amazingly, we had rolled between two good sized boulders.

The kind apartment manager gives us kids and Mom a lift to the airport, while Dad flies out of the field and meets us. In a takeoff over rough ground with very little distance in which to raise up speed, a light load is tantamount.

From the airport we regroup and begin the next leg homeward. Again stopped by a thick cloud cover, we land at Robinson, Connecticut in a grass strip devoid of people, where we sit on a telephone pole next to the runway, tired, hungry and dirty, listening to our stomachs rumble, and waiting for several hours to takeoff once again for Tew-Mac.

At last back on familiar ground at Tew-Mac, we realize just how lucky we are when we see a plane at the end of the runway upside down. Dad coins a phrase and he uses it from now on every time we land.

"Cheated death again," he says with a sinister smile.

Jessica was facing a crisis. She came over one night in hysterics, telling Rip and Ken that her marriage was over. Ripley felt helpless, and comforted her as best she could, and Ken helped too, putting cold cloths on Jessica's forehead while she poured out her problems. When the divorce became final, she moved with her children to her mother's home in South Berwick, Maine.

She and Ripley kept in touch via the mail and phone, but even so they missed each other terribly. Rip planned a cross-country flight to visit Jessica.

"I can't wait to see you," Jess told Ripley that morning on the phone. "How will I know when you land at the airport?"

CHAPTER SIX

"Tell you what, we'll pretend we are part of a James Bond thriller -- I'll fly low over your house as a signal! Then you can drive to the airport and pick me up!"

"How will you know which house is mine from the air?"

"Lay some sheets flat on the lawn, I'll be able to see those. When you hear my plane engines cut, you'll know it's me."

"Cut? What do you mean, cut?" Jess imagined the plane engine, once roaring steadily, silent.

Ripley laughed and explained, "Don't worry, it's a difference in the amount of throttle, you can hear the engine change."

Reassured, Jess spread white sheets in the field behind her house and waited impatiently. When a small plane flew over and she heard the engine noise change slightly then return to normal, she knew it was Rip.

Jess was so excited to see her, she forgot about retrieving the sheets, jumped into her car, and drove straight to Sanford airport. Stopping her car with a jolt, she went inside the airport office, where she found an employee talking to Ripley over the radio. After listening to their exchange and realizing Rip was about to land, Jessica stepped outside to watch and hear the squeak of rubber that signified tires making contact with pavement.

Ripley taxied off the runway to an area set aside for transient plane tie-downs, pulled the throttle out so the engine idled, switched off the radios, and pulled the lean knob out, shutting off the fuel supply. The engine stopped and after a turn or two more, so did the propeller. She then flipped the magneto switch to the off position, and shut off the master.

Jess ran out to meet her, her purse bobbing crazily off her shoulder. She helped Ripley tie down the wings and place the yellow and white chalks on the front and back side of the rear wheel to keep the wind from moving the tail. Ripley removed her purse and sweater from inside the plane, and arm in arm the two walked toward the car, passing the small terminal building.

Ripley suddenly pulled Jess around the corner up against the side of the building, out of sight and earshot of anyone else. She wore a distressed look on her face.

"Jess, you're the only one I can say this to," she began secretively.

"Oh, God, "Jess thought dismally, "she's got a fatal illness or something."

Rip finished breathlessly with, "How do I sound on the radio?"

SEVEN

Between going to school, the airport, visiting with friends and relatives and going to church, the times when we have a day to spend leisurely at home are few and far between. When we do have an "at home" day, I want to do my own thing without any interruption. Today is one of those days, and I am in the living room playing "Make me Dizzy," a game of standing at the center of the multicolored braided rug, humming a little ditty and beginning slowly; then picking up speed, running around the oval braids leading to the outside of the rug until I fall into a heap.

"Go change into dress clothes, we have to go to the airport *right now* for an interview," Mom unexpectedly announces, poking her head into the room before rushing off to her own bedroom to change.

I am angry with her for the interruption, so I stomp through the house to my bedroom where I don a summer dress, white socks and dress shoes, glad that at least Janice and I don't have to wear matching outfits this time. My sister and I, though two years apart, have many identical outfits in our wardrobes. Sometimes the colors vary, but the style is the same. Mom even occasionally has a matching outfit for herself! Trips and visits, the most likely time of family photographs, show us in matching jumpers over white blouses, or frilly dotted swiss white dresses with puff-up slips under them and a red sash with a fake rose at our waist. Even our haircuts are the same.

We don't look the same, though. Janice has Dad's blond hair and grey blue eyes, Mom's square nose and a generous mouth and smile from some other relative. I, on the other hand, have Mom's brown and auburn highlighted hair, root beer brown eyes, Grandma's thin librarian lips, and my fathers straight nose. Janice's face is heart shaped and mine is oval like Mom's.

Jeff and I share looks; he is brown haired and brown eyed, also with Dad's straight nose. He builds models and mini bikes, and someday he will build cars. He can always fix the lawn mower, eat the most dessert, act the most formidable. Jeff also shares Grandma Miller's placid disposition, while Janice received Mom's upbeat sense of humor and sociable personality.

I am not sure which family member my temper comes from, which today I show in full force. I whip my play clothes across the room, thinking about all the fun things I had planned to do. Now I am angry at the newspaper, the reporter, in fact angry at the whole, rotten, dirty world. I stomp my way to the car, and shut the door hard. This will show her, the sourpuss on my face says.

It's hot at the airport, with barely a breath of wind, and shimmers of heat rise from the tarmac. We pose on or around a Cherokee Six, and have to stay in position until the photographer is done. Sweat begins to trickle down my back. "Look this way, smile. Imagine you are going on a fun trip," the photographer tries. I send him a scowl. My dress is sticking to me, and I can't imagine that away.

CHAPTER SEVEN

Mom is eternally patient, not wanting a scene to unfold in front of anyone. I avoid eye contact with her as long as possible, but at one point as I stand on the wing of the Tew-Mac plane, and Mom is in the door of the cockpit, we look directly at each other. The moment stands still as she smiles sympathetically at me. My hard little heart melts, and I return to loving my mom.
The photographer has his picture.

What began for the reporter as an article on a novel hobby for a local housewife quickly became an avenue for Ripley to educate the general public; and she loved to tell about it. She recognized there was still a need to promote the sport of flying. "The Millers feel that a lot more people could enjoy flying if only they could be exposed. They recommend taking the family for a passenger ride some Sunday afternoon at Tew-Mac."[1]

Her outgoing personality, natural good looks, and uncommon pastime gave reporters plenty to work with. It was believed that since flying was so unusual, the pilot must also be unusual. Did her whole life reflect the hobby she participated in? What made her different?

This article reflects public interest regarding women pilots and their personal lives and roles in suburban culture in the early 1960s.

Ripley Miller: Burlington's Only Aviatrix: Says Flying Beats Housework Any Day

Most suburban housewives carry pictures of their children in their wallets, and Ripley Miller does too, but she also carries a picture of the plane she and her husband own, right next to it. Ever since she started learning over a year ago, Ripley has been completely thrilled with her flying exploits. "It's a different world," she enthuses. "Once I'm over the treetops I'm smiling, and I forget all my troubles."

Ripley, who flies out of the Tew-Mac Airport in Tewksbury, became interested in flying because of her husband, Kenneth, a civil engineer. After he got his (pilots) license he began to use flying to facilitate his work. Now Ripley helps him in his work by flying with him. He is planning to cut a hole in the floor of their plane so that he can photograph sites with an aerial camera. While he photographs, Ripley will man the plane controls because; "He trusts me more with the plane than with the camera." The four place Cessna is equipped with "His" and "Her" controls.

Ripley, an attractive 28 year old brunette who looks like a model, believes in "keeping femininity in flying." Although flying is still predominantly a man's world, Ripley feels strongly about maintaining her womanliness while flying. Although she does wear slacks sometimes while flying, she often wears "suits and heels when I'm flying to a meeting." ...As soon as she is eligible, Ripley is anxious to join the Eastern Chapter of the Ninety-Nines, Inc., a world wide organization of (licensed women pilots)... Ripley already attends these meetings as a guest.

How about airplane thrills? So far Ripley hasn't had any scares and doesn't expect them in the future. She feels that flying is completely safe as long as you respect the rules and the weather. Her greatest sensation came from a hundred mile trip she flew to Twin Mountain in New Hampshire. As far as danger goes she feels she takes her greatest chances "getting to the airport."

For a beginner, Ripley advises learning in a Piper Cub. Most of her instruction takes place on weekends. Her enthusiasm is so great that she says, "If I don't fly for a week, I get grouchy!"

Flying has changed the whole living pattern for the Miller family. At first, Ripley admits, the children (Jeffrey, 8, Janice, 6, Julie, 4) were bored

CHAPTER SEVEN

while their parents took flying lessons. Now the children, who show no fear of flying, look forward to the increased number of family trips. Jeffrey recently flew with his father for a hunting trip. In the summer they plan to fly to Martha's Vineyard... Just as other children do in the family station wagon, the Miller children often fall asleep on long trips.

Socially the Millers have made many new friends since they started flying. Much of their time is also devoted to reading about it and "hanger flying." As a housewife, Ripley finds that "I can get my housework done by 9 if I am going flying."

Ripley, who admits to being "a bit of an individualist who likes to keep active" has interests other than flying...[but] flying has so completely absorbed Ripley and her family that she muses, "What did we do before we flew?"[2]

A quick progression of events had catapulted the world into the air age. Wilbur and Orville Wright completed their first flight at Kitty Hawk, North Carolina, in December 1903.

Harriet Quimby became the first woman in the United States to receive a pilot's license in 1911, and also the first woman to fly the English Channel in 1912. A regular United States Airmail service was begun in 1918.

Only seventeen years following the Wright's successful flight, passenger service had begun between London, England and Paris, France, and the following year between Havana, Cuba and Key West, Florida.Records for speed, distance and endurance were set at a rapid pace by both women and men from all over the world. Charles Lindbergh made his historic solo nonstop transatlantic flight in 1927 in a single-engine monoplane.

Amelia Earhart was the first woman to cross the Atlantic as part of a team in 1928, and shortly thereafter was the first woman to traverse the United States solo from east to west and then back again. In 1932, she crossed the Atlantic solo, and in 1935 was the first pilot to make a solo flight over the Pacific Ocean. She wrote about her exploits and her hopes for women pilots and believed that in the future credit would be given to women for their achievements, and their sex would no longer be an issue.

But thirty-three years later, accomplishment was still taking a back seat to gender. Just as Ripley was fascinated with the pilot when she read about the postage stamp fly-away, people were now fascinated with her; how does this busy wife and mother cope with the day to day responsibilities of raising a family and having a flying career?

Of course she cut corners. She established a routine and stuck to it. Sunday lunch was tuna sandwiches, Friday night, hot dogs and beans. Laundry was done twice a week, the bathroom cleaned on Monday, and the kitchen floor on Tuesday. Every Saturday night, tubs and hair washes for us kids. When Janice and I had long hair, it was curled using big pink plastic curlers that had snap on covers.

She didn't fill in information in one-of-a-kind preprinted baby books, or keep our school papers in a scrapbook. Photos at special events were taken by Kenneth or her mother. Birthday parties? "Why give them a party they won't remember anyway?"

She found time to volunteer for a week long Vacation Bible School for two summers, and knitted capes for Janice and me, blue and brown with pompoms hanging from the ties around the neck. And she taught us how to sew on her Singer Touch and Sew, a state of the art machine which she felt lucky to purchase new.

She made her life easier by keeping her daily routines to a minimum. She wore little or no makeup, and what she did wear she applied utilizing the new

CHAPTER SEVEN 47

cultural necessity, the makeup mirror. Her nails were short and unpainted, and other than a penchant she had for shoes, her only "extravagance" was visiting the beauty parlor once a week to have her hair washed, set and teased into a beehive hairdo. But it didn't always last a week.

High on her closet shelf atop eerie white styrofoam heads sat a shag wig and a curled style wig which she could don at any given moment. Janice and I drew faces on the heads in magic markers, forever altering their appearance from ethereal to ghastly. "If your hair looks good, no matter what you are wearing, even old ragged jeans, you will look good," Ripley once advised.

The interest continued, and so did the interviews, once and sometimes twice a year. But promotion is more than just newspaper articles and radio shows. Ripley prepared and taught an aerospace education package to local elementary schools, and represented Tew-Mac Aviation at FBO (Fixed Base Operator) meetings. She attended Winterfest '66, a culture and art show being held at the War Memorial Auditorium in Boston. Sponsored by the Cultural Foundation of Boston, Inc., it was the first event for this organization, which promoted activities in the realm of culture and art.

The Burlington News photographed Ripley, Janice and me, half in and half out of the cockpit of a Cessna Skyhawk. Warren Hupper had brought in two planes for display in the exhibition booth, and asked her to attend, perhaps to show that flying was not limited as a man's sport only. More likely it was because he now recognized her promotion potential.

With insight, she continued to try to advance the sport of flying. Working with the Television Casting Service, Inc., of New York, Ripley tried out for a Sucrets® commercial, which, once made, would highlight women in unusual occupations.

She sent a picture of herself standing by a Cessna 172, and in return received a rejection letter: "I'm sorry that things didn't work out for you with Sucrets®. The advertising agency took forever making a choice for the two spots that they did do and that was two less than they had originally meant to do. Well that's this business. I hope things go well for you and that you always fly in good weather. Kit Carter."

Failure did not bring Ripley down; it strengthened her, gave her resolve. She was growing, maturing as a woman, as a pilot and in her promoting skills.

Ripley summed up her involvement in promoting aviation with the following: "I have appeared on many radio talk shows for major Boston stations and WLLH in Lowell, promoting general aviation and women's role in aviation. I have been the subject of numerous newspaper articles in the *Boston Globe, Boston Herald Traveler, The Lowell Sun, the Merrimack Valley Advertiser, Woburn Daily Times* and the *Wilmington Town Crier;* which have served to spotlight women in aviation on a national level and in the local community."[3]

She always threw in a little of her sense of humor when being interviewed. "No, we don't wear helmets and parachutes any longer. I might invest in a long white scarf, though."[4]

EIGHT

On long car trips Mom teaches us how to sing "You are My Sunshine" and "Swing Low, Sweet Chariot" in a round, and we sing it over and over until she feels we have it just right. She sings to us songs once popular, and I can remember lines but never the entire song; "...if you wore a tulip, a bright yellow tulip, and I wore a big red rose...," and "...would you like to swing on a star, carry moonbeams home in a jar..." or "...mares eat oats and does eat oats and little lambs eat ivy..."

Mom has a myriad of records to play, from Herb Alpert & the Tijuana Brass, to crooning Eddie Arnold's yodeling cattle call song. We are one of the first in the neighborhood to own an am/fm stereo radio with a record player. The unit is concealed inside a dark wood case as big as a bureau, decoratively finished with colonial white knobs on *faux drawers.* She enjoys **Fiddler on the Roof,** *and Herschel Bernardi singing "Sunrise, Sunset" brings tears to her eyes.*

We own a white Volkswagen Beetle at the same time Disney releases its movie **The Love Bug,** *a story about a white Beetle that has an inner life and personality. The little car helps its owner to solve a crime and find true love. Our car has an inner life too, created by three kids crammed into the rear seat. Janice and Jeff wrestle and fight the minute no one is watching; the car is anything but a love bug.*

Whump, whump whump. *The car is bouncing funny, but it isn't my siblings today -- yet. We stop on the side of Route 3A in Burlington center with a flat tire. A well-meaning passerby has stopped to help Mom change the tire. In truth, she is watching as he does the work.*

We had to kill time at the airport while Mom had a meeting with other flying ladies, and now, with the extra delay in the car, we are bored and restless. Mom tells us to sing something together, but we can't decide on the song. Instead, Jeff and Janice try to punch each other by reaching around me -- the middle seater. Jeff leans over me and swings at Janice and she leaps away, and then leans over to swing back. The little car, up on a jack, is rocking back and forth. Once, twice, three times Mom pokes her head into the car and says **knock it off!**

No wonder so many pictures of the three of us show me in the middle. Mom is trying to keep those two apart. But today she is unsuccessful and we discover why she especially likes the Volkswagen; while driving she can reach misbehaving children and pinch their inner thighs.

The history of the Ninety-Nines Inc., the international organization of licensed women pilots, is a chronicle of success. In 1929, twenty-six women, including Amelia Earhart, met at Curtiss Field, Long Island, NY to form a club. Their purpose:

CHAPTER EIGHT

"To coordinate the interests and efforts of women in the aviation field."[5]

Letters were sent out to all known licensed female pilots, numbering one hundred and seventeen. Original suggestions for the name of the Ninety-Nines were typically gender related, and quickly dismissed. By the deadline date, 99 had responded and so due to Amelia Earhart's suggestion, the name for the organization came into being. Among the charter members were women from the United States, England, Australia and Germany.

By 1958, membership had grown to over 1200 women. Geographic regions had been designated and were presided over by governors. Each section was divided again further into chapters and directed by chairmen.

The New England Section became official in 1941, dividing into the Eastern New England and Northern New England Chapters in 1962.

Ripley couldn't wait to be able to join the Ninety-Nines; the only requirement was that she had to be a licensed pilot. Wasting no time, she flew to her first Eastern New England Chapter meeting one month after she had her license.

Monthly chapter meetings featured a film, tour or a guest speaker, and were held at or near airports so the members could fly in. At Hanscom Field in Bedford, Massachusetts, Ripley -- along with thirty-six other women -- toured a four-motored military transport C-124 Globemaster with Air Force Pilot "Chip" Collins.

The entire nose section of this plane opened for cargo loading, only on this day it was filled with Ninety-Nine members plus the crew of the C-124. A photographer caught a great shot of everyone in the yawning cavern of the plane.

Over lunch they planned races, fundraisers and events, introduced new members and caught up on all the latest flying and personal news. Through the Ninety-Nines she met other women like herself, women who loved to fly.

Millie Doremus, a petite woman with blonde hair and blue eyes, was one of the first ladies Ripley met. Millie in turn introduced her to Lois Auchterlonie. Together the three of them flew to the next meeting, where they were photographed by the local paper seated inside Lois' plane.

Lois, a former World War II WASP pilot was a favorite speaker at Ninety-Nine meetings. Lois' recollections, humorous and sobering, reminded the ladies how far they had come.

When the United States entered World War II, an increased demand for labor, coupled with the transfer of men to military service, led many women into occupations formerly held by men. National need and the collective sense of patriotism outpaced social stigma for the working woman.

In September of 1942 the WAFS (Women's Auxiliary Ferry Squadron) and the WFTD (Women's Flying Training Detachment) were formed and operating independently of each other. They later joined to form the WASP's (Women Auxiliary Service Pilot), performing non-combat missions such as ferrying aircraft, courier missions, test flights, search flights and towing targets for training, thereby freeing male pilots for combat missions.

Once the military had enough trained male pilots, the auxiliary force was unnecessary, and the WASP group was deactivated. One thousand and seventy-four women were trained, and thirty-seven lost their lives while on duty.[6]

From one pilot to another they could understand, accept, and relate to one another with admiration, camaraderie, and support.

For others, the perception could be different. Women flying had more to accomplish than just

CHAPTER EIGHT

learning to fly. They were a novelty. A minority. Alternately treated with respect, with condescension, as sex objects, as a joke.

Those who perceived women pilots as they actually were -- intelligent, hard working, grounded yet adventurous -- gave individuals and the industry a boost.

"If airlines overcame public reluctance to fly by placing a few pretty hostesses aboard, think what thousands of pretty pilots are doing to overcome public resistance to General Aviation. Go Ninety-Nines!"[7] As a group the Ninety-Nines worked hard to promote aviation, educate, create opportunities, and generate positive press.

Annual international conventions for the Ninety-Nines met these goals. In 1969 the Ninety-Nines held their fortieth anniversary annual convention at the Waldorf Astoria Hotel in New York.

Governor Nelson A. Rockefeller sent a welcome letter to them, writing, "...You can be proud of the gains the Ninety-Nines have made in furthering the role of women in aviation and in promoting world understanding. I am certain that many of those achievements can be attributed to the unique gifts and resources that women traditionally bring to their commitment to public service."[8]

Their public service aviation programs included scholarships, airmarking, rescue flights, education, safety and awards programs, and establishing a museum and the Forest of Friendship. These activities sought to increase public awareness, form international relationships, and establish a force available in times of need.

Airmarking, a program of identifying airports, was begun in 1933 through the Bureau of Air Commerce. By 1935 the program was executed by Ninety-Nines. Especially in rural areas, where no easily identifiable markings were on the ground, a painted rooftop led the way. Runways or the rooftop of a barn or factory were painted with the airport designation, a compass rose or a directional arrow, to aid pilots in navigation.

Pilots take off into the wind to help give the plane lift, and land into the wind to slow it down, determining the active runway. Without radio contact, a bright orange windsock served as the indicator for the pilot to know the prevailing wind at the ground level. Large painted number designations identify the runway; a big benefit if the pilot's compass isn't working properly!

During World War II, airmarkings located within 100 miles of the coastline were removed as required by the Federal Government. Using black paint, Ninety-Nine members covered the location arrows and airport designations. Since gasoline was being rationed, they used bicycles to get to and from the airport. When World War II was over, the question of replacing the airmarkings arose.

A 1941 photograph of the Northampton Massachusetts Airport designation shows an arrow painted in white on a barn roof. A subsequent photograph shows that arrow being covered in black paint by Ninety-Nines.

Sometime after World War II it was repainted, and again in November of 1966 -- this time the name of the airport in ten foot high letters on its newly paved runway surface. The paint came courtesy of the state highway department, and the painters, including Ripley, Millie and Lois, came courtesy of the Ninety-Nines.

There was a lot more to the Ninety-Nines than public service. Air races, annual proficiency training, education and fundraising served to train, challenge, unite the members and provide opportunities to enjoy flying. To keep the organization growing in a positive direction required countless hours of dedicated

CHAPTER EIGHT

volunteers. And of course they needed to raise funds for races to purchase prizes, hire speakers for meetings, print newsletters and give scholarships; but mostly they wanted an excuse to fly. One of Ripley's favorite fundraisers was the "poker run," because she could bring all of us.

Groups of pilots and planes would meet at a designated airport, then make a scheduled series of flights to four additional airports, each time landing and receiving a playing card. Each landing would have us kids calling out to be shown the newest card. The anticipation of grouping all five into a winning hand was endless. At the fifth stop, pilots would compare and declare their poker hand.

The flight time between airports was about a half an hour long, and the nearly continuous climbing and turning, a short cross-country, then turning and descending was another practice opportunity. There was as always, a time set aside at the terminus to have a bite to eat, and plenty of flying conversation.

All of us children who had joined our moms and sometimes dads for the day, over time began to recognize and remember each other, forming a fragile bond joined by the thread of being part of the Ninety-Nines flying family.

NINE

Ripley and Millie Doremus, 1966 AWNEAR. Undocumented photographer.

Flying at Tew-Mac is an experience for everyone. The runway is only 2,900 feet, so it is used by pilots from other local airports to practice short field takeoffs and landings. Obstacles surround it that pilot and passengers alike take notice of. To the north, a lowland where over the years more than one small plane overshoots the runway on landing and bogs down, noses over or flips. To the east and west, gargantuan pine trees that Dad worries over every time there is a storm; and to the south, Route 38, traffic wires and the mini-golf course. The phrase, "If you can fly at Tew-Mac, you can fly anywhere," is well known to local pilots.

The mini golf is a great spot to aim for on approach, because it stands out with its bright colors and unusual

CHAPTER NINE

shapes, which become increasingly defined the closer you get. Whether it is a bright green dinosaur or a yellow and brown spotted giraffe, the course always has a large creature looming over it that takes on a life of its own as the plane descends to the airport.

At Tew-Mac we rent a tie down for the Cessna 170 under those huge and worrisome pine trees next to the taxi way, and here we spent a lot of time as Mom and Dad prepare for trips or care for the plane. Dad builds a table between two pine trees using a couple of planks; our benches are sawed off logs, and we picnic and play cards and games.

Mom explains the plane instruments and functions to us, while Dad teaches us how to make a proper knot when tying the plane down, how to sump gas from the tanks to remove any condensation, and how to always, always, stay clear of the propeller, whether moving or still. While most families attend ball games or go camping, we hang out at the airport, watching (and critiquing) other planes making takeoffs, landings, touch and go's, and go-arounds.

Practice was a necessity, and the gals of the New England Ninety-Nines recognized this. They planned and sponsored an annual air race for housewives and working women to participate in -- one day of racing between several local airports.

The AWNEAR (All Woman New England Air Race), conducted in cooperation with the FAA (Federal Aviation Administration), ran approximately 275 miles under VFR (Visual Flight Rules). It was a scaled-down version of the AWTAR (All Women Transcontinental Air Race), well known by its nickname, the Powder Puff Derby.

Unlike the Powder Puff, where racers flew from the west to east coast, the AWNEAR was a "round robin" -- a cross-country course that began and ended at the same airport. Speed was not the winning factor; this was a competition in knowing the airplane. The object was to meet predetermined airspeed and fuel consumption rather than finish first. Millie asked Ripley to be the copilot in her Cessna 150 for the 1965 AWNEAR as they left a monthly chapter meeting. Ripley jumped at the opportunity. Even though they were a generation apart, the two women had become close.

The AWNEAR challenged their navigational skills and Millie's knowledge of her plane. The route was a secret until Rip and Millie arrived for registration and impounded the plane. The secrecy gave out of state pilots an equal chance.

The contestants passed over selected airports used as checkpoints in a particular manner, and were identified by ground observers. Failure to pass a checkpoint correctly automatically disqualified the plane.

Ripley was in her element -- flying, competing, and being in the company of other pilots. She and Millie didn't place; but they learned a lot, had tons of fun, and made plans together for next year's AWNEAR.

As the year passed Ripley began to prepare as PIC (Pilot In Command) of the Cessna 170. Several days before the event Rip practiced climbs, letdowns, and low passes.

To get the word out that it wasn't just a man's sport, Ripley arranged promotional articles to be written in two local papers.

The day before the race, Rip and Millie flew to Burlington, Vermont, registered and had the plane impounded; then spent some time relaxing at their hotel and doing a little shopping. They purchased matching light blue plaid dresses and white sweaters to wear as their team uniform.

The jumpsuits and helmets of the 30s gave way to fashionable dresses in the 50s, when the look-alike trend began. The 60s ushered in business suits or

CHAPTER NINE

miniskirts, and for the 70s, pantsuits. Teams may have had matching outfits specially designed for their race, and often would match to the colors of their plane.[9]

The day of the race Ripley and Millie were well rested, dressed alike and raring to go. It was windy, blowing their hair around during the team photo, and making Ripley frown with a crease in her forehead.

They flew from Burlington, VT to Montpelier, VT, Laconia, NH, and on to Keene, NH in choppy air that took their seat out from under them more than once. From Keene they flew to Rutland -- the exact location of the airport eluding them. They made short passes and larger circles trying to locate the airport, losing time and wasting fuel; but at last Millie spotted the strip. They finished the fly over and headed back to Burlington.

Again they didn't place. Tired after fighting the wind in the air as well as on the ground, they were glad to relax and enjoy the race terminus activities -- one of which was a frozen banana daiquiri with dinner. Perhaps two.

TEN

 Mom is getting the hang of traveling with us by air, so we fly all of our "day trips" and "Sunday drives." She soon feels it is time for a longer trip. We leave Dad at home, and excitedly start out for Aunt Sara's in Minnesota; making it only as far as Albany, NY when our radio transmitter dies and we return to Tew-Mac. A "yippee!" to "oh" day.
 Our second attempt begins a week later once the radio is repaired and weather patterns are good again. My brother gets to ride in the copilot seat and help with the maps. Our route has us stopping for fuel at Rochester, NY, Bowling Green, OH, Porter Co. Valparaiso, IN, then Flying Cloud in Minneapolis, MN. Thirteen hours, ten minutes flying out with a one-night stay at Valparaiso. The flight is telling on her; she is shaking her head and pinching her cheeks to stay alert, and her fanny is numb from sitting

CHAPTER TEN

in one position. From Aunt Sara's, she calls Dad, admitting she isn't superwoman. They plan to meet in Detroit so he can help pilot the Cessna 170 home.

On the return trip, we travel from Flying Cloud to Griffith, IN, and then to Detroit Metro where we find Dad, who arrived on a jet that morning. I am glad to see him, except it means now I have to double buckle with Janice again, and lose my window seat to Jeff.

"We need to take a break before flying home," I hear Mom tell Dad, "and bring them to a museum with a lot of walking, to work off some of their energy." Yay! No more flying today. Off we go to the big Henry Ford Museum, with acres of indoor and outdoor exhibits.

Later when pulling in the hotel parking lot, Dad drives beneath the overhanging branches of the one tree, only to see a sign that reads "Birds - Don't Park."

Dad says, "How many birds can there be?" and parks there anyway.

Full with a dinner of take-out sandwiches, we yank on our suits and check out the pool, built half inside the building and half outside. Water whirls around my fingertips as I swim toward my mother, submerged up to the smile on her lips, relaxing at last.

The next morning, well rested and finished with breakfast from the hotel dining room, we troop out to the rental car. We all have a good laugh at Dad's expense; the car is covered, completely white. Every bird in Detroit must nest in that solitary tree.

A newspaper article written previously reported "His" and "Hers" controls in the Cessna 170; the writer was referring to words typed out on a Dymo label maker, then stuck to the wheel. The "His" designation being on the pilot's side, or left seat, and the "Hers" on the copilot's, or right seat.

Even after Ripley became a licensed pilot and they interchanged sides as needed, the tiny labels remained. They were a reminder, perhaps a little joke, of the way they started out flying together. Rip quickly surpassed Ken in flying skills and abilities, and he never caught up.

It wasn't long before Ripley knew that she wanted to do more than fly for fun. She began working as a secretary at Tew-Mac Airport to raise funds for her flying habit. She kept books, answered the phone, organized files and manned the two-way radio to respond to incoming pilots regarding traffic, wind conditions or the active runway. However, secretarial work was not what she had in mind. To fly cost money; to fly as a commercial pilot enabled the pilot to fly *and* make money, and that became her next goal.

To accumulate hours toward the 200 minimum required for a commercial license, Ripley flew fifteen cross-country flights in 1965 and twenty-three in 1966. This was the best way for a new pilot to build up air time, and practice preparation, navigation, and landing at unknown airports.

In manning the controls while her husband took aerial photographs in the rear of the plane from the hatch he'd cut and configured, she accumulated more time from twenty-five photo trips, and together they checked an additional six job sites from the air.

She flew an air race, a poker run fundraiser and several family outings for lunch.

Many small airports had a café or diner to service both fly-in and drive-in customers. At Barnstable Airport in Hyannis, MA, there was the Neptune Room at the airport and Mildred's Chowder House at the end of runway 6. "Sandwiches Made to Fly" at The Airport Restaurant at Barnes Airport, Westfield, MA or The Green Meadow Lodge in Sterling, MA. For picnic lovers, Katama Air Park on Martha's Vineyard Island, where a plane could taxi right up to the beach. For dairy cones loaded with jimmies, the diner at Tew-Mac in Tewksbury.

CHAPTER TEN

Ripley tirelessly continued to build up airtime by flying all over New England to Providence RI, Bridgeport and Danbury CT, Taunton and New Bedford MA, and Portland, ME. She immersed herself in flying with the Ninety-Nines to chapter and section meetings held in Worcester and Mansfield MA, Laconia and Manchester NH, and Sanford, ME. If she couldn't afford to fly one week, she spent time talking about it or planning for it, and by year's end she had 205 hours.

Meanwhile, Millie easily talked Ripley into being her running mate as Vice-Chairman of the Eastern New England Ninety-Nines Chapter offices. Millie had already paid her dues as Treasurer, Secretary, and Vice-Chairman. Now running for the position of Chairman, she wanted Ripley by her side. Like bread and butter, they worked together well.

Nineteen-sixty seven found her back with Warren Hupper in dual instruction learning spirals, 720 degree turns, chandelles and lazy eights in preparation for her test. She practiced steep turns, turns about a point, and wheel landings. By now she and Warren had become easy friends, sharing their mutual love of flying; she had grown used to his affable personality, and he with her perfectionism.

On April 30th, two years after getting her private license, and with 240 hours, Ripley passed her commercial test and completed training to be an advanced ground instructor. At last she could earn money while flying.

That June she began flying passenger rides, scenic flights, photo flights and charters for Tew-Mac. The two month wait may have been due to personal issues, or perhaps a lack of flying work. However, a recommendation letter written years later by Virginia Bonesteel, then Chairman of the Eastern New England Chapter of the Ninety-Nines, told part of the story this way:

"...Ripley's application outlines her accomplishments, but it tells nothing of her persistence in her goal to be an active flight instructor. She discovered, as do many women in aviation, that her FBO (Fixed Base Operator) found her most useful keeping books and running the office..."

And he did. "She was a real asset to the company and extremely bright in handling the business end of the airport," Warren said.

Ripley only needed to be patient, because when the time came and they were both ready, he eased her into commercial flight. Some considered Warren progressive for having a woman pilot on his staff, and others refused to fly with her. Warren remembered one older man climbing out of the plane in anger when he realized the pilot was going to be a woman. Ripley just sent her disarming smile his way.

Her flying work with Tew-Mac started out with a flight to Norwood, Massachusetts for parts and one scenic in the Cessna 172. A scenic was a short fifteen-minute flight carrying thrill seekers. In July she flew five scenics, and picking up business, in August she flew thirty-two -- with fifteen of them in one day!

As soon as Ripley was comfortable in her new role, she began to assert her opinions regarding the atmosphere at the airport. "Warren, when I lead passengers to the waiting airplane we have to pass by the maintenance area. Can you do something about those pictures of naked women plastered on toolboxes and hanging in the guise of a calendar?"

"Well, I don't know if I can," Warren started to explain. "The maintenance area is under separate management."

"You can talk to them," Ripley insisted. "It's bad for business!"

The next day when Ripley reported for work, there was a message from one of the mechanics about a plane being repaired. Written on the chalkboard for employees and customers alike to see was: "6705 Whiskey -- Radio NFG." It's easy to imagine what the worker meant.

She spun on her heel, slapped open the door, and within minutes was out on the tarmac hunting down Warren. Dragging him into the office, she pointed to the board and said, "Not appropriate, Warren."

He had to think fast for an answer this time, and was pleased with what he came up with. "That means Non-Functioning Gyro. What did you think it meant, Rip?" They both had a good laugh at his quick wit.

But the incident stuck in Warren's memory. He didn't want to offend anyone; times were changing, and as a business man he had to change along with them. If women were going to be flying, he'd just as soon have their business at his airport.

Two years later in 1969, there were "more than a dozen ladies taking flying lessons at the Tewksbury school," wrote Ann Geib. "Their main reason being a challenge and the desire to learn how to fly," said Warren Hupper, owner of the school. "Some own their own aircraft, and all of the female students are from Massachusetts."[10]

Half of those students took lessons from Ripley.

ELEVEN

Mom loves Christmas, and believes in celebrating the good old American traditional way -- with materialism. The space under the tree remains bare until Christmas morning when we awake to find endless piles of wrapped presents. But before we run out to the living room and examine the delights under the tree, we find on our beds a stocking stuffed to overflowing with all sorts of delightful treasures. Mom spends a lot of time searching out just the right small toys and goodies for each of us. A pink rubber ball, a little doll in a locket that smells of rose-scented perfume, or playing cards in a leather case from the Deerskin Trading Post.

Always, there is an orange in the toe, which Jeff left in his stocking one year and the following Christmas, we all had a screech of delight on seeing the wrinkled brown prune that rolled out! A tube of Chapstick®, a shiny penny, a Lifesaver® Book of candy.When we were younger, a

CHAPTER ELEVEN

helping of cereal wrapped up to munch on so that Mom and Dad could sleep a few minutes more without hearing the cries of "I'm hungry!"

She has treats for herself at Christmas, too. Red raspberries and red raspberry sherbert, and each Christmastime she fills a red glass candy bowl with raspberry hard candies that have a chewy center. They are made of sugar and will do in a pinch but I prefer her favorite -- raspberry cream chocolates -- which she rarely shares even though I plead, beg and give her my biggest sad eyes.

Our trees are usually like Charlie Brown's tree; more a sign of the times than of our prosperity or lack of it. When Dad was a boy, Grandpa drilled holes in empty spaces in the tree trunk, and inserted limbs sawed off from another part in order to have a symmetrical tree. He tells us this story every year! We just use lots and lots of tinsel to cover the bare spots.

We spend the afternoon at my Aunt Caroline's, or sometimes at Gramma's, where mayhem reigns when you add my seven cousins to us three. The boys hide in a corner doing who knows what with my brother Jeff, while the girls are put to work setting the "kids" table, my youngest cousin, Andy, tagging along. Mom, Aunt Caroline and Gramma, chatting away, put together our Christmas meal.

Not long after Christmas, Mom begins complaining of the cold weather. She says she is feeling pale, and decides with Dad to take us all on a trip to sunny Florida. We plan to visit the Ringling Brothers Circus Hall of Fame, and the Ringling Art and Circus Museum -- as well as stay in a hotel with a pool, see pink flamingoes and toucans at a wildlife center, and go to the beach.

Dad flies at first as PIC until we reach our refueling stop at Dulles, Washington DC. Mom and Dad share the flying, but Mom isn't keeping track of her copiloting time. Instead, she is putting out the small fires brewing in the rear seat from us three kids strapped in for four hours!

For the second leg, Mom takes over as PIC, flying to Myrtle Beach, South Carolina in an additional 3 hours and

45 minutes, where we RON (remain overnight). I squeeze into a hotel double with my brother and sister on either side of me, and early the next morning we continue on to Sarasota, Florida. It's a long flight even though Mom brings puzzle books and travel games to occupy us. When we at last reach our final destination and burst out of the plane, Jeff, Janice and I frantically play tag.

Our first activity is to go to a beach. Mom loves to sunbathe, wearing a modest one-piece suit when there are other people around, but when by herself or with us, she wears a bikini. But Dad can't find the beach and eventually pulls the rental over to the side of the road, where we can see water glistening beyond the trees and brush. Bushwhacking our way through we find a spit of sand along the bay.

Mom throws a blanket on the ground and strips to her bikini. Janice and I wear our new matching red, white and blue swimsuits, with white rubber bathing caps buckled under our chins. But the water looks icky, and we play in the sand. Our fair northern skin is a prime target for the Florida sun; and Jeff and I sunburn, stinging with pain for the rest of the day.

Six months of the year had passed with Ripley in training; now it was time for some flying for herself, for the Ninety-Nines, and for work.

She attended a Ninety-Nines meeting in Providence, and in June, the annual convention held in Washington, DC, making a stop along the way to see her sister Sara and her husband Bob at their new home.

Ripley took the Cessna 170, landing at a little airstrip near Aberdeen. There Sara met her with her two year old twins, Janet and Kristina, and they took photographs of the girls happily clambering around inside the plane.

Back at work Ripley flew a charter to Hartford, Connecticut, two photo flights, and ten cross-country

CHAPTER ELEVEN

flights which included family trips to Plum Island, Katama, and the lakes region of New Hampshire.

Over the summer Ripley and Ken had met Dot and George Anderson, newcomers to Tew-Mac. Ripley asked Dot, a petite woman with dark brown hair, to be her copilot for the upcoming AWNEAR. It wasn't being held until late fall, so Dot and Ripley both had plenty of time to prepare.

The cloudless October morning dawned with a crispness to the air, and fall foliage was at its peak. The two ladies, dressed in a team uniform of a light brown skirt, white blouse, dark brown blazer and brown loafers, flew the Cessna 170 to Norwood, MA. Ripley tucked a Kleenex into her shirt sleeve before they left because the cool fall air left her normally red nose a little drippy.

Ken and George drove out to meet them and take photographs of their departure, return, and the score board with the contestants and flight times listed.

Together with the other 23 contestants they flew to Jaffrey, NH, No. Adams, MA, Johnny Cake (Burlington), CT, Westerly, RI, and then returned to Norwood for a total of 300 miles.

Ripley had done well predetermining her speed, but she burned less fuel than she had calculated. The time away from home was refreshing for them both, and even though they did not place, the beauty of a New England fall day made their efforts worthwhile. Ripley wasn't disappointed -- she had accomplished a lot this year and looked forward to the next step in her plan.

Her goals were becoming more defined as she wasted no time training for her instructors rating. Each rating she gained was a step closer to a higher goal, that of airline transport pilot.

There still were no female airline pilots, and here the salaries were the highest. She was confident she could reach this goal and as income trickled in she purchased more lessons... but would the industry be ready to hire her?

She began practicing short field and soft field takeoffs and landings, stalls, slow flight, straight and level flight, chandelles, eights on pylon and two- and three-turn spins.

When it was time for her test, she flew Tew-Mac's Cherokee along the New England Coast to Portland. She used the time on the flight to mentally go over a few last points, and wasn't surprised when she easily passed. But she could barely contain herself as she returned to Tew-Mac.

It was a cold and clear, breezy, January winter day. The landscape broke away in jagged pieces from the Atlantic Ocean where squiggly lines of white told her the waves were being whipped to whitecaps. The thrill of passing the test and reaching another step in her goals was reflected in the beauty of the day. She spontaneously broke into song:

Summer and winter and springtime and harvest, sun, moon and stars in their courses above, join with all nature in manifold witness... Her strong, alto voice sang out.

Ripley now had 307 hours flying time, and was ready to teach. She told Bob Morris of the *Wilmington Town Crier*, "I love flying and I love teaching, so I guess it was only natural that I should get to be a flight instructor."[11]

Her nonchalant way of describing her job to a local reporter did not come close to the passion she felt about teaching people how to fly. Kenneth described her enthusiasm this way, "She got so excited to teach a new student and see them catch on. It was like putting a shovel full of coal on the fire and feeling the satisfaction when it lights."

TWELVE

As a family in the sixties in suburban America, we have a fairly typical and predictable home life. It is when we leave our home that we are unlike anyone else.

Our plane is old and has its own funny odor of musty upholstery and aviation fuel. Janice and I buckle into the same seat belt and Jeff squeezes in next to us, because the Cessna 170 that Mom and Dad bought has only four seats.

Strapped into place in the tiny back seat, we watch over Mom's shoulder as she snaps the master switch on, turns the ignition key which activates the magneto switch and generates electricity for spark, primes the engine, flips open the window and shouts "clear" (a warning to anyone standing nearby), and lastly pushes in the starter button.

The prop turns with a jerk, then kicks in and speeds up as the engine happily vibrates the entire plane and its occupants.

Once running, the radios are turned on and a mixture of hums, whines and voices crackle out from the speaker so loudly it hurts our ears. Rolling clumsily to the end of the taxiway to the start of the runway, the plane is given its final check -- a run up.

All the flight controls are checked for freedom of movement. We look out the window at the wings and watch as the ailerons, sections on the back edge of the wing that control turns in the air, are lifted up and down by the wheel. Then Mom pulls a lever mounted on the floor to check the flaps, also located on the wing but these move into several locking positions, and enable the plane to maintain flight at slower speeds.

The rudder, a movable section on the vertical portion of the tail, is controlled by the lower half of the foot pedals, and serves to turn the plane in the air or on the ground. And the elevators, similar to the ailerons only located on the back of the tail section, are controlled by pushing the wheel in or out. These control the attitude of the nose of the plane while in flight: pull back on the wheel and the nose comes up, push in and the nose drops.

Time to check the engine and its oil pressure by running up the RPMs (pushing in the throttle) while holding the brakes firmly. The roaring engine strains to move the plane forward, but the brakes hold it back.

By now the anticipation of a flight is beginning to build in me. My stomach is getting butterflies, and my heart is going a little faster than usual. We all look at the sky into the landing pattern for any incoming traffic and when there is none, move onto the runway.

"Tew-Mac traffic, 1857 Charlie preparing for takeoff, runway 3," Mom says into the hand-held microphone.

Pushing in the throttle, we gather speed, the plane trying to wander to the left or right and Mom maneuvering it into a straight line using the foot pedals. Engine noise drones louder and vibration becomes stronger as we pick up speed. I look down at the pavement rushing past, staring without blinking, and feel rather than see the plane pick itself up off the ground, defying gravity. Like a theme park ride, we soar.

In the air every trip is different. On a still day, the ride is like weightlessness as we glide along. In the winter we wear our coats, hats, and mittens to stay warm, breathing on the glass and drawing pictures in the frost left behind. Some days we are tossed around by turbulence that lifts us up, drops us back down, and leaves our stomachs somewhere in the middle. Airsickness doesn't go away until you have a long nap, and the smell of throwup in a sic-sac bag on a three-hour flight remains with you forever.

Let's take a plane...anywhere!

Some families swim or water ski together. Others play tennis or even scuba dive. Many picnic or travel together, but how many "flying" families do you know? Where most children remember car vacations, the Miller children remember flying trips. The ability to fly their own plane has provided the Miller's with vacation flexibility..."[12]

Ripley shared her flying enthusiasm and need to practice with spending quality time with us, flying to Nantucket Island or to Jaffrey, NH, where we could walk across the street to the Silver Ranch and enjoy an hour of horse back riding, which the ladies of this family loved to do.

"What do the Miller children think of hopping in the family Cessna single-engine plane as a means of transportation? 'Well,' said Mrs. Miller, 'they get in the back seat just as they would in a car. After a while, the thrill wore off a little, and the children think of it as another way to get to a destination.' "[13]

Tew-Mac was a home away from home for us. Late one evening, Ripley heard Jeff howl as he stepped on a pencil and the lead point drove into the bottom of his bare foot and broke off. Unable to remove the point, Ripley had to make a choice between an emergency room visit and the bill that would accompany it, or an alternative. She remembered there was a poker game going on at the airport, and that one of the players was a doctor.

Jeff made the choice for her. Terrified of hospitals since he had a head wound and stitches to close it, she knew he'd take his chances on the game table.

The cigarette smoke in the air swirled when Ripley opened the door and led in a hopping Jeff, explaining his plight to the surprised men. Moments later, the point safely out, the game resumed as she led Jeff back out again.

Ripley's passion for flying spilled over into her household where she organized events with the Ninety-Nines, read flying books, collected and wore jewelry of airplanes and birds in flight. She had pins, earrings, and necklaces of the Ninety-Nine insignia, which was designed for the Ninety-Nines by Louis Tiffany, American artist, designer and friend of Amelia Earhart. In the living room hung two relief metal casts of biplanes, on the desk sat a wire sculpture of a bird. She also kept a World War I leather aviator helmet with green goggles, and her silky long white scarf.

When we moved in 1968 from our ranch house in Burlington to a cape in Wilmington, Ripley replaced the living room furniture. She loved to see the color red in her decorating, and chose a colonial theme, with red fabrics and a dominating pattern of eagles in the material and matching wall prints.

The independent spirit in her admired Richard Bach's book, *Jonathan Livingston Seagull,* and consequently she loved all objects that showed

CHAPTER TWELVE

seagulls in flight. Her youthful nature adored Charles Schultz's *Peanuts* character Snoopy, in his aviation persona, the World War I Flying Ace. She had a Snoopy doll complete with aviator helmet and scarf, a pin of him dressed the same way sitting on his doghouse, the "Sopwith Camel," and Schultz's book *Snoopy and the Red Baron*.

Her no-nonsense attitude toward life was best displayed by her comment "If a person is out in cold weather and is feeling discomfort from cold, they are not dressed properly." Her serious side enjoyed biographies of pilots Charles Lindbergh, Sheila Scott and Eddie Rickenbacker, who she admired because he had a strong faith in God. In her copy of the book, *I Must Fly, Adventures of a Woman Pilot*, by Sheila Scott, Ripley underlined passages where she felt akin to Ms. Scott, such as never feeling alone in the sky and being more fulfilled from flying than from anything else she had ever done.

All of the facets of her personality came together when she was training, teaching, or flying. One of her favorite pictures was of Amelia Earhart in her flying jumpsuit, leather helmet and a string of pearls. Ripley said that picture represented the type of woman she wanted to portray: "Strong and knowledgeable with a definite feminine side."

That strength was needed when she received the news that her sister Sara's daughter, Janet, was ill with leukemia. The disease had struck quietly, and swiftly took her five year-old life from the family that loved her.

Kenneth and Ripley broke the news to Jeff and Janice, and then to me, and the next day they took a jet to Washington, DC to be with Sara and Bob, attend Janet's funeral, and help with little Kristina. They couldn't stay as long as they wanted; we still needed their attention, and work could only be delayed for so long.

Ripley's heart was torn as she left; between the ache of knowing she was needed there, and the desire to be home with us, each one so much more precious to her now.

THIRTEEN

Photo by Pigeon. The Lowell Sun, October 6, 1969

So Mom doesn't have to pay a baby sitter, she takes me with her to the airport on Saturdays. Warren's daughter Jenny is the same age as me, and we keep each other busy -- and from being a nuisance to our parents. We are the airport (b)rats.

Tew-Mac is family owned and operated, and the small, friendly atmosphere allows everyone to know everyone else. We talk to all the pilots and passengers, hang on the fence outside watching planes and people come and go, chew gum and give a burial service once the flavor is gone.

We play until we are bored, then beg for change to buy drinks from the old soda machine. There is orange, root beer, or Coke® in glass bottles with a cap that has to be pried off using the bottle opener set within the machine. The top drops with a clink into a square metal cup inside, and Warren opens the door with a key to give us all the caps to play with. Begging for change for the soda machine is an art. "Just one thin dime, it's so thin you won't even miss it."

When the airport office is remodeled, we have a room upstairs to ourselves most of the time. The entire second floor is one long room, with tables, chairs, books and maps, and eventually holds a flight simulator, too. Upstairs in the relative quiet we draw and color at the tables or write on the chalk board. This big room is usually where Mom finds us in between her flights, and sometimes either she or another instructor shoos us out so they can conduct a ground school lesson.

Today Rex Trailer landed in his helicopter! He is the star of his own TV show, **Boom Town**. Each week his show features a special guest, songs, his horse Goldrush, and comedy with one of his sidekicks, Pablo or Cactus Pete. My favorite is the show about the kids from North Reading who race sled dogs. Wait, Mom must know Rex Trailer -- they both fly -- maybe she can get me an autograph and I can show it off at school!

Well, if not then maybe she can take us up for a flight, and we will plead for the roller coaster ride of a stall. Our appetites are insatiable for the swell upward, watching for the loss of vision to the horizon, the straining sound of the engine trying to maintain flight, and the difficulty controlling the plane. Then comes the anticipated sudden drop of the nose toward the ground, rush of the plane back to speed, and the thrilling lurch of the stomach. Over and over she practices while we crow, "Again!" or "Do another!"

CHAPTER THIRTEEN

"So you want to learn to fly? Well, the general qualifications don't require that you have the physical coordination of an athlete, or that you have to see like an eagle. Nor does it state that you need the courage of the Apollo 11 crew. It does say that you have to pass a very simple medical examination. You are asked to learn basic facts about aircraft, weather, and navigation. One very important thing is that you're a mature and responsible individual,"[14] Ann Geib wrote in the *Lowell Sun*.

The FAA requirements for a private pilots license evolved over the years from ten hours of solo time to forty hours of combined solo/dual time, a medical exam, and a Radiotelephone Operator permit, issued by the FCC (Federal Communication Commission). The license is granted for the lifetime of the pilot provided they pass a medical exam by a doctor and a BFR (Biennial Flight Review), administered by an instructor every two years.

Over the course of her first year, Ripley taught various aspects of flight to over twenty-five students, and continued to fly scenics and charters such as one to the State College in Pennsylvania. She flew the family to New Jersey to visit her grandmother, and attended Ninety-Nines meetings in Sterling and Barnes, MA, Hartford, CT, and Newport, RI. By years end she had logged an additional 234 hours flight time.

Her logbooks, brief by necessity, showed the steps made in aviation by her, and now contained her students' progression as well: Larry Benjamin -- *introd. st & lvl*. The steps in teaching someone to fly begin with an introductory flight including straight and level.

Rich Young -- *climbs & glides, orientation*. The student then progresses to medium turns, climbs and glides, climbing and gliding turns, steep turns, slow flight, and beginning navigation.

On to stalls, an action in the air where the pilot causes the plane to "stall" or to stop flying. The engine does not stop running, but the speed of the airplane combined with the attitude (angle) of the nose and wings cause the lift to weight ratio to no longer support the plane. The reaction of the airplane is simply to "fall" forward, the nose drops and the lift increases as the wing attitude returns to a point where lift can support weight. Usually the stomach is left higher in the air! Making the plane stall is preliminary to landing, as this principal has the plane touch ground just as it stops flying, or stalls.

In her second year of flight instruction Ripley taught forty-four different students, supervised the solos of seven, and checked out eight others. Pilots wishing to fly a different model plane, for instance from a single engine Cessna to a single engine Piper, must be checked out in the new plane.

In addition, Ripley flew fourteen photo flights with representatives from Cole Photo, Aladdin Studios, Les Vant Photo Service and of course, her husband. There were thirty scenic flights, seven shuttles and one package delivery into Logan airport, a charter flight to No. Springfield, VT, a flight for parts, and trips for Ninety-Nine meetings to Newport, RI, Manchester, NH and the 40th annual convention in New York City. Family trips were fewer this year, only one to Plum Island, and a two day visit to the Washington area to see Sara, Bob and Kristina.

She also took her father up flying for a local scenic. She had always made the dutiful required visits on Christmas and birthdays, and invited her parents as well as Ken's over for Thanksgiving dinner. Still, it was a shock to her to realize that her alcoholic father looked every day of 85, instead of the 65 year old he was. She returned home that day trying to remember everything good about him, pushing away any bad memories.

CHAPTER THIRTEEN

She thought back to high school, when all she wanted to do was move out and start her own life. He had given her an ancient baby blue Chevy convertible for her graduation, which she tooled around in during the year she was single. By the time her family was started she didn't have the money to keep up with repairs. Getting rid of that car was one of hardest things she'd ever done.

He died unexpectedly from pneumonia in January of 1970. She grieved the knowledge that the relationship could now never be whole, and endured the sadness of finality.

It was a cold winter, and the wetland areas of Wilmington had frozen solid. Looking to lift her spirits Kenneth suggested an afternoon of ice skating. They skated long enough for fingers and toes to be numb, but the promise of hot cocoa at days end made the wait to thaw bearable.

Ripley was feeling confident by this time even though she did not have a lot of experience skating. A push the wrong way, a bit of imbalance, and her entire body was flung onto the rock hard ice. A loud "bong" echoed across the bog as the back of her head whiplashed to the ice. That ended the skating for the day, in fact, they never went again. She didn't get a concussion, but from the sound Ken was sure she had, and he watched her anxiously for several days.

She adored the attention he gave her, but it was flying that consoled her, renewed her. "I forget my troubles when I'm above the clouds," she said.

FOURTEEN

Ripley with brother-in-law Al Fay, after flying into Logan in the twin.

The loud hum of the engine coupled with our passage through the air makes such a din that we have to shout in order to be heard. Silence is required during takeoffs and landings, so Mom can concentrate and hear any radio messages. Headphones for the pilot and copilot of our plane come much later, and there are never any for the passengers.

With a window seat Janice and Jeff can watch the countryside unfold beneath us, but I am getting bored, and start elbowing them for more room. A little disagreement breaks out. Mom is busy flying and ignores us; Dad looks over his shoulder and tells my brother and sister to move over some. Naah, naah!

CHAPTER FOURTEEN

Finally! Mom points out Lake Amberjesus to us and we fly a low pass over the shoreline, pulling the throttle in and out so the engine makes a revving sound. This is our signal to Uncle Carleton and Aunt Neva who are waiting for our arrival. Leaving the lake behind, we land at a small airport in Millinocket, Maine, tumble out and wait for Uncle Carleton to pick us up.

The day is turning hot, heat bugs begin to buzz, and the promise of cold northern waters under enormous blue skies is more than my patience can bear. At last! My uncle arrives and now it is an agonizingly long half-hour drive to the lake.

The water is low this year and boulders of all sizes pop up from below the surface. We jump from one to another for quite a distance before the water becomes deep enough to swim in.

What is left of the day passes by too quickly, and disappointed, I pack up my things. When our plane leaves the airport for the return trip, it is already dark, and the air is cooler and still. Worn out, the three of us in the back seat keep our thoughts to ourselves.

Pockets of lights appear below, indicating suburban clusters. Portland, ME and Portsmouth, NH show clearly to the east, and farther off and to the west, Lawrence and Lowell, MA. The headlights of cars look like a coil of Christmas lights, stretching out along I-95.

The lights in the cockpit glow red, so as not to spoil Mom and Dad's night vision when looking outside. It is a good thing, too, because there have been a few times when night vision was badly needed, and tonight is one of them. The runway lights at Tew-Mac do not automatically turn on once darkness sets in. Instead, we radio ahead to a specified frequency, and that triggers the lights to come on. Tonight we cannot make contact and are forced to make a landing in the dark.

Each of the lucky four with a window seat is supposed to peer outside and shout out what they can see. Relying on the street lights that illuminate Route 38, our location is

accurately pinpointed and we begin our descent to the runway, which is a darker patch of black pavement within a dark patch of grass that surrounds it.
I hold my breath and cross my fingers.

While training others, she continued her own training, flying the Piper Apache to build up time towards her multi-engine rating. She learned single-engine procedures for a multi-engine aircraft, emergency and instrument procedures. On her thirteenth wedding anniversary, she met all FAA standards for a multi-engine rating. "Once you start taking flying lessons, you get kind of hooked!"[15] she said.

She could now fly a plane with more power over longer distances, carrying more weight. It gave her the opportunity to take on additional tasks at Tew-Mac, such as the Boston Air Taxi, a shuttle service to Logan airport. Shuttles to Logan were popular until greater landing fees were imposed -- making it financially prohibitive for small airplanes -- and the service stopped.

The job of flight instructor and commercial pilot was certainly varied and interesting; it needed no additional thrills from bad weather. Six months passed and Ripley was once again reviewing basic airwork, this time for her instrument, a rating that allowed her to fly solely on instruments in certain weather conditions.

Her new instructor, Eddie Burke, was a dark-haired fellow with a gleaming smile who led her on weekly lessons. She was "under the hood" -- a plastic visor that limited vision to the instrument panel -- learning VOR approaches (Very high frequency Omni Receiver), ILS approaches (Instrument Landing System), missed approaches, ADF orientations (Automatic Direction Finder), holding patterns, tracking, and interceptions.

CHAPTER FOURTEEN

The twofold test procedure had the student pass a written before taking the flight test. Gossip at the airport centered around the difficulty of the written test, how many times each person had attempted it, who had failed, and who had barely passed. With fear of failure looming over her, tempered with the confidence of study time, Ripley took the written and passed with flying colors.

She was quoted in the Lowell Sun as saying, "Today, quite a few boys are taking flying lessons but the girls just don't seem aware that they can learn."[16]

Although the curiosity of a woman pilot was beginning to wear off, the number of women in flying related jobs continued to be low.

In 1969 women with their private pilot certificate numbered 8,755 nationwide; with their commercial license 1,470 -- and in all of New England, Ripley was one of only a dozen female flight instructors.[17]

In fact she continued to be the only female instructor at Tew-Mac, and taught twenty-three students in 1970, for a total of 282 hours, compared with the 238 hours she taught the year before. She soloed two students, gave recommendation rides for two students, and checked out three pilots.

Add to this ten scenics, three photo flights, and another long distance trip to New Jersey, where she took her grandmother for a ride in the Cessna 170.

Ripley preferred teaching young adults, claiming they were more at ease in the plane, concentrated better and were more adaptable. But she thought women were still behind in aviation and the space program. She hoped that the college-aged girls would be the ones to "make it."

There were inequities even though huge social changes occurred from 1960 to 1970 in our country.

In 1963, Betty Friedan's *Feminine Mystique* challenged long established ideas of womens' roles.

The Equal Pay Act was signed into law, and President John F. Kennedy's Commission published its results of the first government sponsored study on the status of women.

A Civil Rights Bill introduced in 1964 outlawing employment discrimination based on race, color, religion or national origin, was subsequently put into law with the addition of a tiny word: sex. It was added because some Congressmen believed that if you included women, the Civil Rights legislation would not be seriously considered. They were wrong.

NOW, the National Organization for Women, began in 1966 and their initial work started the notion that men were equally responsible with women for the care of children and upkeep of the home. In 1970 the birth control pill became widely available, and now women had the ability to plan their child bearing years. The Virginia Slims slogan "You've come a long way, baby" was certainly true. But they still had a long way to go. The Airline Industry as late as 1970 would not employ a stewardess if she had children.

For some, the transition from viewing women in traditional roles to qualified pilots was a big jump. Bob Buck, a Captain at Trans-World Airlines, had this to say in the article "Thank Heaven", published in the official program for the 24th annual Powder Puff Derby:

"It's always been a little difficult to forget that gals are gals and think of them as pilots. This is only natural for the male especially if the male is fond of gals, and I am. But despite my normal interest in females as females I have found, over a long period of years, that females can really fly...

"But seriously, there isn't any reason why women who want to fly professionally cannot be airline pilots. They have done wonderful flying jobs in instruction, taxi work and most all other parts of the profession. So why not the big job, too?"[18]

CHAPTER FOURTEEN

The idea that a woman in the cockpit was upsetting to passengers continued to linger. At the end of 1970 there were 70 women with airline transport licenses -- although none worked as commercial airline pilots, a few were copilots on smaller airlines.

Encouraged by the changes in our country on the whole, and discouraged by the strong emotions of hatred and anger brought about in some people by these changes, Ripley sought to maintain a middle ground. She celebrated the differences between the sexes, maintaining her femininity, while proving she could participate on the same level as a man in the sport of aviation.

While groups such as NOW, WEAL (Women's Equity Action League, the more conservative offshoot of NOW), the National Federation of Business and Professional Women, Federally Employed Women (FEW), and the YWCA fought against sex discrimination on a national political level, individuals like Ripley and her piloting friends determined to change their own course and the future course of aviation, by learning, teaching, promoting and leading others to the best of their ability.

FIFTEEN

Ripley, Janice and Mona Budding, Worcester Air Show. 1970

    One unexpected benefit from dragging us to church every Sunday is that Mom begins to place her book-learned knowledge of Christianity at a place where it is meant to be -- in her heart. And Dad feels the same way. So one Sunday at the Concord River in Billerica, the two of them are baptized along with several other people.
    We park in a nearby lot, and with twenty or so witnesses including us three children, walk down an overgrown embankment to a dirt shore where the darkened blue-brown waters of the Concord ripple. I am surprised to see them wade waist high into the water in their street clothes! One by one they profess their faith, are fully

immersed and drawn up again. "Now wash me and I shall be whiter than snow..." we sing.

Mom sees to it that we never skip Sunday School or church. We attend summer vacation Bible school and winter retreats, pot luck lunches, kite contests, church picnics, and baby showers with other families from church. Mom initiates bedtime family prayers, where all five of us gather on our knees around one bed and take turns praying aloud. Janice makes a mistake in her prayer one night and I laugh out loud. Mom yells at me; I'll never make the same faux pas again. Or as Mom calls it, fox pass.

"Welcome to the exciting, busy world of the woman pilot!"[19] began the preface in the cookbook authored and produced by the Eastern Chapter of the Ninety-Nines in 1969 when Ripley was ending her second term as Vice-Chairman, and beginning her first term as Chairman. Aptly named *Wings in the Kitchen, A Lady Pilot's New England Sampler*, the book contained easy-to-prepare traditional New England recipes. It also contained quips and quotes on flying, and brief descriptions of some favorite airports and their restaurants in Maine, Massachusetts, New Hampshire and Connecticut.

The opening page contained an imaginary dialogue between a control tower and pilot.

"Tower: Niner Niner November Echo, do you wish to declare an emergency at this time?

"99NE: Negative, but request a straight-in approach. Company arriving for dinner in 30 minutes.

"Tower: Roger, understand 99NE, straight-in approach. Suggest Beef 'N Noodles in a Hurry, page 42, for dinner, at your discretion."[20]

This cook book was available at a Ninety-Nines informational table set up during the 1970 air show held at Worcester Municipal Airport. Posters depicting the history of the Ninety-Nines and photographs of the founders hung as a backdrop; pamphlets, brochures and tokens were spread out on the table. Ripley, as Chapter Chairman, along with Janice, stood available as representatives to answer any questions.

Ripley was dedicated, highly organized, gregarious and indefatigable, a combination guaranteed to enable a busy woman to accomplish much. She spent many hours arranging for and chairing the monthly meetings, initiated having an opening prayer, and confidently led the chapter members through business. She attended special events as an envoy, planned fundraisers, contacted members, arranged press time.

As a born again Christian, with the maturity she had gained in marriage and motherhood, personal loss and tragedy, she was truly heartfelt about her faith and her love for Jesus Christ. She told her mother-in-law she wasn't afraid. "Flying makes me feel real close to God; it is something wonderful up so high." She had discovered her inner peace.

And that peace empowered her and radiated to family, neighbors, friends, pilots and reporters. "I myself have been searching for the same inner peace that she had found," wrote Sue Haselmann, a fellow pilot.

Ripley took naturally to her leadership role. She was one of the lucky few who had found the hobby that gave her great joy, turned that hobby into a career, turned that career into a passion, and then turned that passion into leadership to help others along the way.

Joanie Burley, a new pilot, was so affected by her first meeting with Ripley that she later wrote the following: "The first Ninety-Nines meeting I attended was during a planning session at Worcester for the (annual) convention. I decided that day I would join. My decision was based almost entirely on the

CHAPTER FIFTEEN

incomparable professional way that Ripley was running the meeting.

"I quite truthfully had never imagined that there could be a woman with her competence, poise, and personality. After seeing her more, I have only thought more highly of her and become more and more impressed by her and her many capabilities and accomplishments.

"For me, Ripley epitomizes many of my goals both in aviation accomplishments and contributions, and also in personal development. I'm afraid I can never really hope to equal her, but it's good to have an example and proof that there is a reality for which to be able to strive."

Ripley was also a gifted hostess, planning many of the monthly meetings, and several were held at our home in Wilmington. She genuinely enjoyed having company, setting out her avocado-green fondue pot filled with melted chocolate, surrounded by wooden handled skewers and bowls of fresh fruit.

One cold January night in 1970 a new member was "pinned" into the Ninety-Nines. Lillian Emerson began flying as a passenger with her husband, then after being criticized for flying without any training herself and having six young children, enrolled in the "pinch hitters" course offered by the AOPA. She was taught from ground school and a little airtime, learning such basics as radio use, who to call in an emergency, and the fundamentals of flight. She found she enjoyed flying enough to continue training.

Months later while on the last leg of her three leg solo cross-country, Lillian began to have second thoughts. *What are you doing?* she wondered, *You should be home sewing.* And after a brief moment she thought in reply, *Wait -- you don't even like to sew!*

One month after gaining her private license, Lillian drove to Ripley's home for her pinning ceremony. "Ripley," Lillian said, "lit up the room when she came in. I held her in awe. Here she was, a wife, mother, pilot, leader in the Ninety-Nines, and younger than myself. But I knew she had fought hard for what she had achieved."

Asked if she had ever encountered any prejudices while flying, Lillian replied, "Not prejudices, but sometimes when my daughter Chrissie and I flew together we would get a double take as we stepped out of the plane." Her daughter soloed at age 16, licensed at 16 1/2, and was "pinned" at age 17 by Amelia Earhart's sister, Muriel Morrissey.

Together they formed a mother-daughter flying team that was not as unusual as it may sound. At a Ninety-Nines convention in Coeur d'Alene, Idaho, there were enough mother-daughter pairs that the committee planned a hike and picnic just for them.

The group of a dozen pairs climbed a small hill that seemed a mountain to Lillian, where the picnic was being spread and sodas cooled in a nearby brook.

In discussion Chrissie mentioned how her mother didn't like flying over the Rockies, and Lillian added, "If you have to go over them to get here, you have to go over them to get back!"

And the laughing reply from the seasoned Ninety-Nine pilots, "We'll teach you how to crash!"

Lillian said, "The ladies of the Ninety-Nines are always prepared for anything."

That spring Ripley came up with the idea of having regular BFOs. At the beginning of summer she set up a flying outing schedule for every Wednesday with Virginia Bonesteel --who sported long black tresses and cat's eye glasses -- and another fun-loving Ninety-Nine pilot, Harriet Fuller. The BFO was her abbreviation for "Bonesteel-Fuller Outing."

Rainy days were canceled, but on CAVU days they flew to Wiscasset for lobsters, Block Island, Nantucket and Martha's Vineyard for the beach, Jaffrey and Twin Mountain for the scenery and more.

CHAPTER FIFTEEN

These trips weren't limited to the three of them; Lillian, Cora, Millie or Jean joined in, too. Not only did Ripley have fun and travel to interesting places, she led her peers into a tight-knit group.

Ripley was as serious about accomplishing tasks for Ninety-Nine events as she was with flying. "One year I was Ripley's roomie for an overnight at a convention," Lillian Emerson recalled, "and she was proud of me for my hard work in losing some excess weight I had gained. She encouraged me for my accomplishment, but boy if you dropped the ball on some job or another, she let you know about it!"

The 1970 annual convention, hosted by the New England Section this year, was held at The Mount Washington Hotel in Bretton Woods, New Hampshire. Each year the host of the event planned activities beyond business meetings, scholarship awards, dinners, and speakers, that highlighted their hometown area. Of course being in one of her favorite areas in New England didn't hurt when it came time to say "Yes" for those volunteer tasks.

Nestled in the White Mountains, with a gorgeous panoramic view of Mt. Washington and the surrounding range, the hotel commanded nearly as much awe as the scenery which unfolded behind it. A long curving driveway led to the massive structure, a turreted, painted white building with a red tile roof, that looked like a fairy tale castle springing up from the books of childhood.

In charge of registration, Ripley was all business and intent on the convention; and entirely dedicated to her volunteer position and the group of ladies for whom she held such affection. Pictures show her at the registration table, on a mountain cable car ride, or seated at a banquet table with other Ninety-Nine Officers.

While listening intently to the scheduled speaker, Ripley licked her bottom lip and caught it with her teeth, holding it a moment and then letting go, a habit that accompanied deep thought. Was she thinking about the speaker's topic, or about this event? There were 300 Ninety-Nine members in attendance, 80 planes tied down at the White Mountain and Twin Mountain Airports, and countless details to be tracked and attended to.

Meanwhile the children, including myself, attended the activities planned for us, wandered all over the beautifully landscaped grounds, and went swimming in the pool or horseback riding on the hotel horses.

Split into boy-girl teams, we played a loud and spirited game of tag, running up and down the wide red carpeted stairs that led from the lobby to the second and third floors of the hotel. Even the uniformed elevator operator got into the act by giving the girls refuge, and transporting us to another floor when the boys started to catch up.

We dressed formally for dinner and ate with the convention members, behaving only because we were at last caught under our parents' watchful and expressive eyes.

SIXTEEN

Billie Downing and Ripley at Tew-Mac. 1971

On the way home from church we pass an ice-cream store, a curse to Mom and Dad and a delight to us kids. To force our parents into stopping, we begin chanting about a mile before we get there. "We want an ice cream! We want an ice cream!" The trick is to start out chanting softly and then rise to a crescendo, hoping the pressure will make them crack. Louder and louder as we approach the store, the thrill and hope and anticipation making us three kids laugh and bounce and love each other for the moment.

Sometimes Dad pulls in and we all shout "hooray" until Mom shushes us; sometimes he drives on by and our disconsolate "ohhs" hang in the air. Do they plan this torture together? Once he slowed down approaching the driveway, and we thought he was turning in. We started hiphooraying, and then he sped back up.

He even once pulled in the u-shaped driveway only to pull back out again!

Today he pulls in and gravel spits out from under the tires as he makes the turn. Jeff, Janice and I fairly chortle with happiness as we trip our way to the order window. Orange sherbert, buttercrunch and maple walnut. My brother and sister and I all taste each others' cone. Mom gets a coffee flavored cone, and I try a small taste. Blech!

Sunday is family day and if we aren't having tuna at home, ice cream at the stand, or going to a friend's house, we go out to eat.

I love this particular restaurant because they have baskets of white, milk and dark chocolate for the customers, though it's never enough for us candy hounds!

Another church family is with us, June and Ed Cobb and their three daughters who are the same age as us.

"You can't eat the whole thing, I bet you a dollar!" Mr. Cobb is saying. Jeff insists he can, and moments later the waitress sets down a dessert known here as the "awful-awful" pie. Layers of cake, chocolate sauce, whipped cream, ice cream, caramel sauce and more. We all wait in anticipation to see who will win the bet. A long, long time later, Jeff admits he can't eat it all.

On Thanksgiving, we go to the Cobbs' house for turkey dinner and Mrs. Cobb makes something I've never tried before -- dumplings. She lets me lift the cover of the pot, steam escaping, to look at the white mounds of dough surrounded by bubbling stew.

Mom recently completed a Wilton cake decorating class, and knows how to make flowers out of frosting, Christmas bells out of sugar, and a cake frosted to look like a basket, covered with her favorite flower, red roses. This basket-of-roses cake sits in the place of honor on the Cobbs' dining-room table along with a pecan pie -- Jeff's favorite -- and an apple pie.

Mom and Mrs. Cobb offer five dollars to us girls if we do all the dishes. Oh boy! Five dollars! But then we see the state of the kitchen with the sink piled high with dirty

CHAPTER SIXTEEN

dishes. Hours later when the last dish is done, we split the money five ways. Hardly seems worth it...

"Fun in the Sun in '71", the forty-second annual international convention, was being held in Wichita, Kansas this year. Ripley rented the Cherokee Six from Tew-Mac -- sharing the expenses, but not the flying -- with Billie Downing, Georgia Pappas and Cora Clark, all three pilots and Ninety-Nine members. Rip wanted concentrated time to practice for her upcoming instrument rating flight test.

Engines humming, sporadic voices and beeps coming over the radio, the group proceeded west. About halfway into the fourteen hour trip, they encountered some mountainous cumulous clouds, and thunderstorms building up on both sides of the route they wanted to take.

Although flying on instruments, but without having the instrument rating as yet, Ripley had to remain faithful to VFR, which required 1000' vertical and 2000' horizontal distance from cloud cover, and three miles of visibility. She knew she could call in to the control tower and request a deviation from their flight plan, or try to sneak between the clouds. The four pilots spotted an opening up ahead.

"It may be a sucker hole," said Billie, a Texan by birth who relocated to New England and learned to fly at Hanscom Air Force base where she worked as a civilian. They could fly through the opening, but once in the hole, the clouds might converge. They might fly closer to the clouds than the regulations allow, or worse, lose their visibility.

Her dark eyes were solemn as she and Rip conferred. Together they decided there was enough room between the clouds.

Ripley maneuvered through the sucker hole while Billie, Georgia and Cora held their collective breaths. Then they encountered more cloud cover. Trapped!

Ripley was forced to radio the control tower and ask for a higher elevation.

She said, "The guys in the control tower are probably saying, 'Those women pilots -- they always want to *see* where they are going!'" The four of them broke out laughing.

When they reached Kansas it opened beneath them in a checkerboard pattern of green, yellow and brown. Fields plowed, growing or laying fallow.

Billie was busy taking aerial photos. An avid photographer, she was building up a collection of 8 X 10 glossies of places where she had flown. From the air, the view became one dimensional, and formed designs that are implausible on the ground. She mused, "Such an interesting pattern, as far as the eye can see."

"Thank God for VOR tracking," voiced Ripley, and everyone nodded in agreement. In eastern New England a local pilot can navigate visually by observing highways and terrain, but in an unfamiliar state with similar terrain, an out-of-town pilot could easily lose their way.

Once landed and checked into the hotel, the women freshened up and made their way to the convention registration desk. Ripley opened her itinerary, scanning the list and in a businesslike way, began calling out items of interest. First to catch her eye was the opportunity to take a lesson in a multi-engine Cessna 401. She wasted no time and had several lessons in the two days there.

The ten and a half hour return flight occurred without a mishap; Ripley was again PIC using only the planes instruments to keep them on course and flying level. Twenty-four hours of instrument flying prepared her well, and she passed the test just nine days before her fifteenth wedding anniversary, on June 21, 1971.

CHAPTER SIXTEEN

Island hopping! Rum swizzle parties! Treasure hunting! Ripley and Kenneth celebrated with 125 contestants gathered on Grand Bahama for the 7th Bahamas Flying Treasure Hunt. Eighteen aerial views of the Islands had to be located and identified with correct latitude and longitude coordinates. First prize: a building lot of land in Stella Maris on Long Island.

Enroute, they remained overnight in Elizabeth City, NC. From West Palm Beach, FL, they flew over the Atlantic to West End, on Grand Bahama, for two days of events and contests while grounded. Then they began the treasure hunt, flying from West End to Marsh Harbor on Great Abaco Island, Rock Sound on Eleuthera Island, and finally Stella Maris where they RON again.

Kenneth said, "The Cessna 170 was the slowest plane there. It took us several days just to get there, and most of the other contestants had already checked out the Islands. It was a real disadvantage for us."

He had the job of searching for the clues, which gave him a headache and a little case of airsickness that lasted the rest of the day.

The next day they continued to Nassau on New Providence Island, the final RON destination where the award ceremonies were held. They shopped at the local tourist stores and spent time at the beach, where the wind blew Ken's toupee up in the air and out of shape.

In the evening they attended a dinner and dance held for the contestants. Ken left their table briefly and when he returned he found his social wife dancing with another man.

"It was just an innocent dance," Ripley said in defense. But Ken was annoyed and the evening soured. Having an early start planned for the following day, they called it a night.

At 8 am and back in the air, they traveled straight through until two the next morning, stopping only to fuel at West Palm Beach, FLA, McKinnon, GA, New Bern, NC, and Atlantic City, NJ. Ripley was PIC for the entire trip, 36 hours in eight days; but both Ken and Rip were equally worn out on their return. They fell into bed exhausted and slept late the following day.

SEVENTEEN

Ripley and Georgia Pappas, 1972 AWNEAR.

Girls day out! Mom, Janice and I are flying to Nantucket Island today. We file a flight plan by phoning in our route, and report in when we arrive. Halfway there, we can no longer see the mainland. A little creepy!

This time I get to be copilot first, and Mom lets me take the wheel once we fly outside of Boston Control area. She has already explained the heading I need to stay on, and the instruments I am supposed to watch. I also have to scan the sky regularly to watch for other planes, and to keep myself from getting vertigo, which can occur without a fixed horizon to look at. On some days, the color of the sky and the color of the ocean are so similar, you cannot see the horizon! Not so today, there is a thin line differentiating the two.

Mom takes over as we approach the island, and I spend our descent time looking at the miniature boats, buildings and little toy cars below us. I marvel at the sparkle of sunlight on the many backyard pools. These scenes always remind me of Mister Rogers' Neighborhood.
It's a perfect summer day to stroll around the town. Already we grow used to the overriding smell of the ocean, and the seagull calls that blend with the voices of other tourists. Window shopping is fun if it includes the promise of buying a souvenir. At a crowded little shop, we go inside and I spend a lot of time looking at each trinket; finally settling on a one-inch high glass bottle stopped with a miniature cork, containing a tiny white sail boat on a painted sea of blue.

In the early spring of 1972, Ripley phoned Ann Geib of the *Lowell Sun*. "Ann, would you be interested in doing another story on a female pilot? This time I am racing in two events, one is a local all woman race, the other international. We are racing from Ottawa to Florida..."

Ann barely let her finish before she gave her enthusiastic reply. Her editor gave the piece front page of the "Design for Women" section of the paper.

The excitement is mounting for Ripley Miller and Georgia Pappas who are competing this Saturday in the All Women's New England Air Race in Windham, Connecticut. Ripley is piloting a 1972 American Traveler which boasts a 150 hp Lycoming engine, and Georgia will be her copilot.

For Ripley, it's her fourth try at the yearly event. Last year, Miss Pappas placed second. The two lady pilots are sponsored by Warren Hupper, owner of the four seater aircraft to be used in the race...

Although the plane is new, Ripley hasn't had any trouble getting accustomed to it. "I find the plane very responsive," said Ripley, "and it beats

CHAPTER SEVENTEEN

the competition for what's available in performance, price, and the way it handles. It's a real joy to fly."

Ripley informed us that during the flight she and the other pilots will be making identification flybys at two or three points. "There's no danger," confirms Ripley, "of a collision with another plane because the planes are not in close formation. You may see one of the planes along the route, but even that is unlikely. You're all at different speeds."

Ripley says the Traveler is a moderately speeded plane. Qualifications (to enter) include having 50 hours of cross-country time. They may solo in the race or have a copilot. All contestants must, of course, be licensed. Both girls are well qualified, Georgia has 800 hours and Ripley has logged 1,600 hours...

...Georgia also has her commercial license and her instrument rating. She has recently been appointed a Safety Counselor by the Federal Aviation Administration Council. Last year, Georgia, who started flying in 1965, won the New England Air Derby. She is employed by Massport at Logan Airport and resides in Arlington...

...Ripley loves the sense of adventure and said her main reason for entering the race was simply for fun. When asked why she decided to have a copilot when it's not mandatory, she replied, "Two heads work better than one, but it's also nice to have company."[21]

In this AWNEAR, Ripley and Georgia flew with 21 other planes from Windham, CT, to Fitchburg, MA, then to Rutland, VT and back to Windham. As in previous races and like many other teams, Georgia and Ripley wore matching outfits which corresponded to their plane colors.

This time it was a pair of white bell-bottomed pants and a belted yellow tunic. Even the accessories tied into the theme; both wore a white belt and shoes but carried a black bag, a subtle reference to the black pinstripe and lettering on the plane.

Posed by the plane before takeoff, big dark sunglasses in place to guard against the brightness of the day, the pair looked young, modern, attractive, and confident.

When the race was over the pair posed again, this time as second place winners. Their average speed was below the projected 109 knots by only .3 knots. Christina Robb, reporter for the *Boston Globe*, positioned the cameraman for a side shot of Ripley. Stepping back, she asked Ripley how it felt to place second. Sitting in the cockpit of the American Traveler, she gave her widest grin and said, "Terrific!"

Billie and Lois had placed first, flying Lois' 180 hp Piper Cherokee. The four ladies hugged and talked and stood for pictures, chatting about landings, speed, plane performance and fuel use.

The *Boston Globe* published Christina's story and the beaming photo of Rip in Sunday's paper. Excitement was high and it didn't let down, because Ripley and Mona Budding, a Ninety-Nine pilot with short blonde hair, a good-natured personality and her own awesome plane, were due in Ottawa in nine days for the start of the Angel Derby.

The All Women's International Race, which began in 1950, was sanctioned by the NAA (National Aeronautic Association) and sponsored jointly this year by the City of Fort Lauderdale and the Florida Women Pilots Association.

Rip was swamped with getting ready, because she had to prepare not only for herself, but for the entire family. Kenneth was flying down commercially to meet her for a short vacation. The house had to be in order and groceries available for her mother-in-law, who didn't drive, to come and watch us. And our clothes and school needs had to be ready for the

CHAPTER SEVENTEEN

week. Mona was in a similar position, as her husband and daughter were flying down along with Kenneth.

Between work, home and packing, Rip called Mona late one night to finalize plans. "Mona, were you able to find a blue blouse?" The two had planned matching outfits, but didn't have time to shop together. So like girls in high school, they matched their wardrobe over the phone.

Mona's multi-engine 337 Skymaster, aptly nicknamed a push-me-pull-you plane because it had a front and rear engine, instead of the standard location of one per wing, was a different craft for Ripley to fly and as copilot, she eagerly anticipated getting her hands on the wheel. With an airspeed around 165 knots it was much faster than her creaky old Cessna 170.

The twenty-second Angel Derby left Ottawa International Airport Monday, May 15 and raced 1900 miles across the Canadian border through six states, to finish May 17 at Executive Airport. The scheduled stops were at Syracuse, NY, Akron-Canton, OH, Louisville, KY, Dyersburg, TN, Montgomery, AL, and Gainesville, FL.

Mona's plane was red, white and blue, and together they wore navy blue polos with red, white and blue skirts when they arrived in Fort Lauderdale, where Kenneth snapped pictures of them. Out of the forty participants who finished the race, Rip and Mona placed twelfth.

At Executive Airport, special events were planned to last several days. Photos show the women holding oranges, wearing white skirts and dark red blouses, standing to the side of a large sign proclaiming, "City of Fort Lauderdale Welcomes Contestants and Friends of the 1972 Angel Derby." For the evening awards ceremony they wore matching white, full-length sleeveless dresses.

Ripley continued her tradition of trying new planes and along with Kenneth, took an hour of acrobatic training each day for three days at Tamiami Airport in a Pitts biplane. They spent some time sun tanning, boating, and relaxing with Mona and her family. With a little bit of a tan, and thoroughly satisfied intellectually and socially, Ripley and Mona took two more days to fly home in the Skymaster, while the others went on ahead in a jet.

The drone of the engine filled the silence between them on the trip home. At length Ripley asked, "I know my mother-in-law will have taken good care of things, but for the one day Kenneth was with the children, do you think the dishes were done?"

EIGHTEEN

Ripley and Jeff, Photo by Wallace. The Lowell Sun, August 29, 1971.

Mom and Dad have found a new place where we can go for supper and satisfy all age groups. Shakey's Pizza in Nashua, NH, has entertainment that is as good as the food. Scents of pizza, popcorn, cigarette smoke and sticky spilled soda mingle together, assaulting my nose at first but which I quickly become accustomed to. A long, dark-paneled waiting room contains a player piano, where for some pocket change you can pick out a tune on the rolls of paper with precut holes, place it in the piano and listen to it play Americana favorites.

In the adjacent dining room with family style tables, a banjo player and pianist lead us in an old time sing-along: "She'll Be Coming Around the Mountain," "I've Been Working on the Railroad," and "Oh Susannah." They play a lot of ragtime, too, like "Alexander's Ragtime Band," one of Mom's favorites, because it reminds her of her Uncle Bud who plays ragtime piano. When the band takes a break, a large movie screen mounted high on the wall plays silent cartoons.

Popcorn is free, drinks are served in large pitchers, and we dive into huge pizzas. The screen on the wall switches from cartoons to the lyrics of songs, but Mom already knows them all. The banjo player starts up again, and we sing along, "...those wedding bells are breaking up that old gang of mine..."

Photo flights were a bit more varied in 1972 as Ripley carried Les Vant and his equipment to Fall River, Sandwich, Boston, Salem, Plymouth and Gloucester. She flew over a gas fire and an SPCA (Society for the Prevention of Cruelty to Animals) location. There were nineteen scenics, one of them a seacoast. She ferried two planes for Tew-Mac, one from Cuyahoga, OH and another from Bangor, ME. For fun she flew her family to the Reading Air Show in Pennsylvania, and made day trips to Warren, VT, and Millinocket and Kennebunk, ME.

The planes she used varied with the flights she made. For flight instruction, she used the Piper PA-28 Cherokee, the American AA-1 Yankee, 1A & 1B Trainers and the Cessna 150. For instrument use, cross-country flights, racing and air taxi use, she flew the American AA-5 Traveler and the Piper PA-32 Cherokee Six, a six seater. For charter flights, vertical photo flights, instrument use and racing, the Cessna 170 and Cessna 172 Skyhawk. On more than one occasion she used her Cessna 170 to give cub scouts free rides; and for one of their trips, she flew the

CHAPTER EIGHTEEN

leaders -- one of them her brother-in-law Al -- into Logan using the Tew-Mac Apache.

She flew the Piper PA-23 Apache for multi-engine and instrument use, the Piper J-3 Cub for training, tried a World War II Stearman biplane, and flew the Pitts biplane and the high wing taildragger Citrabria for acrobatics. Ripley received her glider rating in a Schweizer 2-33; and having one lesson in a helicopter was only a tease to have more.

She kept her flying standards high, receiving the Ninety-Nines APT (Annual Proficiency Training) for four consecutive years. She organized low rental fees through Tew-Mac, and offered her instructor services free for Ninety-Nines who wanted to take the APT.

In addition to flying in the AWNEAR with Georgia and the Angel Derby with Mona, Ripley made a BFO trip to Nantucket, and took Ann Geib of the *Lowell Sun* and Bob Morris of the *Wilmington Town Crier* up for an introductory flight.

She moved up the ladder in volunteer positions in the governing of the Ninety-Nines, holding office as the Chapter Vice-Chairman for two terms, and Chapter Chairman for one term. At one time or another she was Chairman of Section Nominating, Amelia Earhart Scholarship, and Aerospace Education Committees, and a member of Chapter Nominating Committee.

Ripley came to appreciate her training and experience in secretarial skills that she used briefly after high school, and again at Tew-Mac, when she became Secretary of the New England Section.

She was on board with Jean Batchelder as Governor, and Millie as Vice-Governor. Jean, a pilot since 1962 and a member of the Northern New England Chapter, worked for an aviation company as well as wrote an aviation column for the *Manchester Union Leader*. Holding down the fort with the Eastern Chapter were Virginia as Chairman, Harriet as Vice-

Chairman, Georgia as Secretary, and Billie as Treasurer.

Training, practice and hard work paid off not just for Ripley but for all women in aviation. It is now plain to see that ideas regarding women and flying moved along with the times engulfing mainstream American culture. From "...an attractive, 28-year-old brunette who looks like a model"[22] in 1965, to "...when she talks about her work it's with a note of fearlessness..."[23] in 1973, the style of reporting changed along with the ideas.

Some reporters wrote a short story in place of a news report, as this following article by Bob Morris of the *Wilmington Town Crier* shows:

With Rip Miller, It's Flying With A Difference

Rip Miller walked onto the tarmac with a steady pace, eyeing the red American Trainer parked by the hanger with the relaxed gaze of a professional pilot.

Rip's eye glanced westward for a second as a thirty-two year old DeHavilland dipped over Main Street and came to rest on the Tewksbury Airport runway, its maroon body throwing off a slipstream that swept through Rip's long brown hair.

Then Rip got into the cockpit, thought momentarily about the wind, and took out her compact and carefully combed her hair before giving me an introductory flight lesson. (author's note: Ripley would not have stopped to comb her hair, this was the time to pay attention to the machine.)

Flying with Ripley Miller, who is also Mrs. Ken Miller and Rip to friends, is hardly any different from going airborne with any flight instructor, except if you are a man you have to watch your language. Mrs. Miller has all the credentials anyone would need teaching flying, whether single or multi-engine, clear weather or

CHAPTER EIGHTEEN

instrument nightmare. With 1700 hours in the air, it isn't easy to imagine anyone with more self assurance in a mass of flying metal than she.

"It's a good plane," she said as the aircraft cut solidly through the air. Once we had leveled off at 1500 feet she leaned back and took her hands off the controls. The plane moved onward as though it had no need for human beings. No one flies planes by the seat of his pants anymore.

As we didn't want to visit New Hampshire, Mrs. Miller took over control of the plane again and on my verbal prodding, talked of herself while she cruised over the incredibly green fields of Tewksbury at a leisurely pace...

Ripley provided no tales of white knuckled landings on fogged runways, no students bezerk with fear. "You think ahead of what you are doing in this business," she said. "We always check the weather, for instance, before we take-off. As for the students, before they even see me they usually have an idea of how to fly and so forth. I've never been frightened by any students activities."

While some men might be nervous about having a woman teach them how to drive, Mrs. Miller reported no difficulties in teaching flying. "Everybody is too busy studying and learning to be bothered about that," she said. She said about the only thing common among her students who range from teenagers to grandfathers, is their preconceived ideas about flying in light planes. "They usually find it quite different from what they expected," she said.

I was no exception. I expected the plane, which had less power than my economy car, to be buffeted around more by the wind. It seemed as solid in the air as my car felt on the ground.

Mrs. Miller said she had no fear of danger where flying was concerned. "There is probably more danger in driving a car than flying a plane," she said.

However much she loves teaching persons how to fly, Ripley Miller finds the job offers her the added benefit that she only has to work when her three children are attending school. She says her children, Jeffery, 15, Janice, 14, and Julie, 12 are fairly blasé about her airborne activities, as are her mother and octogenarian grandmother.

"The children consider flying just my job," she said. "My mother has gone with me to photograph her farm in Kennebunk, Maine, and I've even had my grandmother up over New Jersey."

To Rip Miller flying is something you feel, like the taste of a good apple, something words can't describe. The best she can say about her business and hobby is, "You are working in another dimension, not confined by roads or gravity. The freedom of flight is a beautiful thing."[24]

NINETEEN

Even with the window open all night my second-floor room remains stuffy. I grab my swimsuit, an extra shirt, and a paperback, heading back downstairs to the relative coolness of the living room to wait. I languish as Mom gathers the rest of the family for another trip, this time to visit Jessica at her cottage on Stearns Pond in Sweden, Maine.

Jeff, Janice and I squeeze into the planes back seat where the air is thick and hot. Thankfully, the flight is short. We buzz over the cottage, drooling at the sight of the blue water under us, anticipating a cool swim, and proceed to the nearby airport to wait for Jessica to come and pick us up.

We wait a long, long time at the airport -- really just a mowed strip in a field with a windsock -- in the sweltering sun with no shelter or a place to get anything to drink. My parents hire a farmer to drive us to the dirt road that leads *to Jessica's cottage. With suitcases in hand we walk the rest of the way. Our hair is plastered to our brows, our throats dry and croaking, when we show up unexpected at Jessica's door.*

Jessica and Mom begin jabbering about how they got their messages mixed up, while we stand first on one foot and then on another, waiting for permission to suit up and swim.

Ripley once again trained, this time for her instrument instructor rating. On January 15, 1973, after spending a number of hours in dual instruction with Eddie Burke, Rip flew to Norwood, Massachusetts in Tew-Mac's American Traveler to have an FAA agent test her for the rating. He signed her logbook *"instrument flight instructor check sat"* (satisfactory). Her teaching level jumped into a new category.

Her aspirations did not stop there. She dreamed of an immediate goal of obtaining her airline transport license, and a distant goal of setting aviation records. She began researching for possible records to attempt and what was required of pilot and plane.

However, she was short on funds again.

Continually paying for lessons drained her income. Plus, she and Kenneth had purchased a hilly piece of land with a huge errant boulder on it that reminded her of her childhood home in Wakefield. Together they had dreamed and sketched, and produced a set of floor plans to build a new house. A bigger kitchen and entertaining area for her, a garage and shop for him. An architect's bill for them both.

She decided to apply for the Ninety-Nines Amelia Earhart Memorial Scholarship, a fund given to help members advance training. She recruited Eastern New England Chapter Chairman Virginia Bonesteel to write a recommendation, while she began laboring over her application letter.

CHAPTER NINETEEN

Late at night at the kitchen table, a snack keeping her company while her family slept, she penned and crossed out and re-penned her letter.

"I wish to advance my career in aviation by achieving the highest professional goal -- Airline Transport Pilot. This advancement will make more readily available future job opportunities as chief pilot of any flight school or aviation college. There are four colleges in the immediate area with active aviation programs, Hawthorne College, the New England Aeronautical Institute, Northeastern University and North Shore Community College.

"I would like to teach both the theoretical and practical aspects of aviation and feel I would be better qualified with an Airline Transport Certificate in addition to my present experience."

Ripley chewed the end of her pen while she dreamed into the future. *Where do I want to be in my career in ten years?* she pondered. The next sentences came easily.

"In five years time my youngest child is graduating from high school, and in the extended future a college of further distance, such as Purdue, might be possible. Other opportunities could include pilot-in-command of transport category aircraft or extended over water flights. Either of these opportunities would be welcomed.

"Through my past experience as an instructor, I have especially enjoyed working with young people, and feel that working with today's youth at college level, promoting and teaching aviation is of prime importance."[25]

The final copy was typed using Kenneth's office typewriter, on two sheets of paper with a carbon sandwiched in between. Now came the hardest part, waiting for a reply.

She eagerly collected the mail each afternoon, flipping through the stack as she walked into the house. The bills were tossed aside in the hope of finding something better. When the envelope with the Ninety-Nine insignia arrived, she stepped into the kitchen and stood next to the counter to excitedly tear it open.

"Dear Mrs. Ripley Miller, Thank you for your application..." She skipped over the formal greeting to the next paragraph.

"We are sorry, but..."

Without reading further she dropped the letter on the countertop, opened the nearby closet, lifted out the carpet sweeper, plugged it in and mechanically started to vacuum.

All the while she was thinking. Thinking of a way to get what she needed (money) to get where she wanted to go (more training). She refused to put off her dreams until the future. Would a second job do? Take Jeff out of the all boys Catholic school? Looking at the budget from all angles, she finally came to a conclusion.

It was time to give the reporter a new story.

TWENTY

The smell of meatloaf hangs in the air as I step into the kitchen and grab a seat at the table. Mom is sitting on a chair at the end of our ridiculously short countertop with the phone glued to her ear. I think she sees me as I pick from the salad waiting for dinnertime, but at second glance I see she is looking over my shoulder as she raptly listens to her caller. Then she lets out a cackle of a laugh that makes me jump and exclaims to her caller, "What a scream!"

I squirm with embarrassment.

Flying jobs were scarce and the demand for them high. Actively looking for a new job while still teaching, Ripley applied with Corporate Air in Hartford. She rented a Tew-Mac plane and flew to Hartford-Brainard for an interview in May, beating out the other applicants and landing a job flying cargo. Ripley was ecstatic even though it required flying the graveyard shift five days a week.

It meant more hours and better pay. She went to Tew-Mac to give Warren her resignation, but when she stepped into the office a wave of nostalgia came over her. Would she miss the seating area with the butt-filled ashtray centered on the beat up coffee table, or the glass-front counter displaying current sectional maps, cross country planners and logbooks for sale? She turned and found Warren watching her, perhaps waiting.

"I got the job, Warren," she said simply.

There was an awkward pause while Warren decided what to say. "Working nights is gonna be tough," he said gently, "especially flying a twin alone." His searching look recognized her mind was already made up.

She jumped into the job with the same gusto she put into all her other tasks. On June 1, 1973 her logbook read, "Tom Diloge, recommendation ride for private"; and on June 4, "BOS, BDL, BOS, BDL, HFD, BOS ". Her job was to fly a cargo of canceled checks twice nightly between Boston, MA, and Bradley International, CT, with a stop at Hartford-Brainard, CT, before returning to Boston. Flying a twin engine Baron, her shift lasted about six hours with actual flying time a little under three hours per night.

In no time at all she became a regular fixture with the night force at the Federal Reserve Bank. Relaxing with a light meal, drinking a cup of coffee to keep her edge through the night, she chatted amicably about flying and family. They formed a special bond with

CHAPTER TWENTY

each other; the ground crew knowing she was somewhere up in the night sky and waiting for her arrival to load or unload, and Ripley knowing her work family waited to greet her.

Meanwhile, the AWNEAR committee had run into a snag. Time was running out, and there weren't enough volunteers. Knowing Ripley could handle almost anything, a plea was given to her to become the Operations Chairman. Unable to say no, she wore herself to a frazzle to please those around her.

Slowly word got out that she no longer taught but flew cargo on a nightshift. Lois Auchterlonie wrote her a note that said: "Congratulations on your new job! How gratifying to know the doors are opening for qualified women pilots!

"Am sure you more than qualify and bet you passed more 'tests' and passed with better score than your competition. What a bonus to be flying a Baron -- it may spoil you for the single engine put-puts! Can't say I envy your constant diet of night flights, but it must be a beautiful sight coming into Boston and Hartford. At least the air should be smoother and less HOT! Best wishes for success in your new position and future endeavors in aviation.

"P.S. Thanks for your effort in Operations at AWNEAR -- you did great & under unusual time conditions!"

With AWNEAR and the initial upheaval of starting a new job over, she breathed a small sigh of relief, but she still had to deal with building a new house with Ken, and having the three of us kids on summer vacation. Keeping the house quiet during the day so she could have undisturbed rest was no easy task. Summer passed by slowly for her, as she bided her time for school to begin again.

TWENTY-ONE

Photo by Al Fay.

Our needs for Mom have subtly changed from supervision, training and physical care; to permission, taxi service and pocket money. It shows in the way we all get along with her, or don't get along. On Saturday Mom drives my friend Lorry and I to a horseback riding lesson at Andover Riding Academy, and later that day she and Janice have a knockdown, drag-out fight over doing the dishes! I come out of my room to sit at the top of the stairs wide-eyed, listening to the yelling. I can hear Janice try to stomp away, Mom tries to stop her, somehow they wrestle a little and Janice drops her glasses and steps on them.

She says, "Dammit, my glasses broke," and the slight swear raises Mom's temper to another level.

CHAPTER TWENTY-ONE

"What did you say?"

By this time Janice is coming around the corner to the bottom of the stairs, clutching her glasses with one hand and the railing with the other, pulling herself up the stairs and breaking away from Mom's grasp. She runs into our room, closes the door, and blares her music.

I sit frozen and desperately try to melt into the wall as Mom looks at me. She hesitates only a split second, climbs the stairs and enters the bedroom. I hear a cracking noise as Mom breaks Janice's favorite record into pieces.

Over the next few days they make amends, both apologizing for their behavior. Mom takes Janice on the job with her on a Sunday when she only has to make one flight to Bradley and back instead of two. Jeff got to go with her a few weeks ago, and I can't wait for my turn.

Janice tells me about the smooth night air, and how they flew above the clouds which spread out below them like a blanket, a fall harvest moon lighting their way. The closeness she felt with Mom by the time they returned. After midnight!

In the darkness on the ride home in the car, "Both Sides Now" by Judy Collins is playing on the radio, and they sing the words together, "...I've looked at clouds from both sides now..."

But Mom has three teenagers, and we carelessly take turns acting our age. I plan a small party for my thirteenth birthday, and Mom and I buy several bottles of otherwise prohibited soda for the event. Just before the party starts I look and see that one of the bottles has been opened and a glass or two taken.

"Who did this?" I screech in disbelief.

"I only took a glass, you have three bottles!" Jeff glowers.

Mom tries to mediate between us, Jeff slams himself in his room, and Mom ends up yelling at everyone. She still has to finish putting together the food and delegates final chores to me. Guests arrive before we finish, we force happy faces, and guess what? There is enough soda to go around.

Once fall arrived and we returned to school, the pressure was off and Ripley's sleeping time was more productive. She still had the weekends to deal with; on these she would remain partially in the night schedule and partially in the day schedule. But things were looking up.

"*Lady Airlines Pilot Undaunted*" reported the newest achievement in Ripley's career path, clearly showing her boost in personal confidence:

> To hear Ripley Miller discuss her job, those who fear flying in airplanes might blanch just a little more. Mrs. Miller, of Wilmington, wife of a civil engineer and mother of three teenage children, is a pilot for Hartford based Corporate Air, Inc., and when she talks about her work it's with a note of fearlessness of the dangers that airplane pilots and passengers can be heir to.
>
> "It really just takes a level head to do what I am doing. You have certain things you need to watch for, but it's not so terribly complicated.
>
> When I am up there, I have a sense of freedom, of working in another dimension. I can just sit and smile."
>
> Though undaunted, she admits that problems can arise. "One no-no," she said, "is running out of gas. You just can't pull over to the side of the road." But she points out that with foresight, most problems can be averted. "You always have to be looking ahead." She said, explaining "You can do that in a preflight (examination of the plane) and a run-up (examination after the engines are on but before takeoff)." Phrases like "preflight" and "run-up" became part of Mrs. Miller's vocabulary about 10 years ago when she joined her husband in his newfound pastime of flying.
>
> She insists that a woman pilot is hardly a rarity these days, though she adds that a woman flying for an airline is. She said her friends and neighbors are not brought up short when they hear

about her job, and to her fellow workers, she says she's "just one of the guys."

Shuttling back and forth between the kitchen and the cockpit at odd hours of the night does not create too many difficulties for her family, she maintains. "I get to sleep around dawn and get up anywhere between noon and 2. That should be about the same time they get home from school," she said. By 9:45 p.m., garbed "usually in pants, sometimes a dress but no long white scarf or goggles," she's on her way to Logan Airport.[26]

Flying into international airports with their complex systems of plane management required Ripley to be in contact by radio with the Approach Tower, the Control Tower and Ground Control as she entered each designated area, and gave her a daily dose of learning experiences. She didn't log any personal flying time once she started with Corporate Air in June, just a six and twelve month check for herself.

TWENTY-TWO

    I have a new craft project for a decorative wall hanging, hammering tiny nails into a preset circular pattern, then wrapping colored string around those nails to form a design. Like a 3-D Spirograph™. Janice and I sit on the floor in our room and begin to hammer. We start out vigorously pounding the nails.
    But Mom is trying to sleep and hollers from her bedroom one floor below for us to stop. We do stop, for a moment, and then we try hammering again, this time v-e-r-y quietly. It makes no difference; every tiny tap amplifies to the turquoise quilted bed below. She charges up the stairs to yell at us, eyes bugging out, face red.

CHAPTER TWENTY-TWO

We know she is serious this time, so outside we go, abandoning the hammer and nails. We try to keep ourselves busy, but we still have to come inside to use the bathroom or get something to eat, and both rooms are next to her bedroom.

Lack of sleep robs Mom of what patience she has. Janice watches her buttering a piece of bread and rudely asks, "Gee Mom, got enough butter?" and Mom bursts into tears and leaves the room.

Rip realized that she was reaching her limit, and she knew it was the night hours taking their toll. She argued with Jeff over buying a motorcycle, with Janice over going out with a boy who wouldn't come to the door to pick her up, with me about piercing my ears. Kenneth went to social engagements alone.

She drove to Tew-Mac to see Warren. They had a little heart to heart, and she confided to him how much the new job was wearing her out. Like the comfort of a heavy blanket, she asked if she could have her old job back.

The next day, September 23, 1973, Corporate Air flight 124 with Ripley as pilot departed Windsor Locks, Connecticut at 2208 e.t.d. (10:08 p.m. estimated time of departure) on a VFR flight plan for Logan International Airport. At 10:25, the visibility and cloud cover was 300' scattered, 1600' broken, 5 miles visibility with fog.

She was advised by Approach Control to expect an ILS (Instrument Landing System) approach in the event of weather deterioration. Sure enough, five minutes later the weather had deteriorated to 300' broken, and Ripley was placed on instrument radar vectors for the approach.

Another twelve minutes and already in the landing pattern for Logan Airport, Ripley was told to "go around" (regain altitude and return to the beginning of the landing pattern). A Delta jet was still on the runway, unable to find the turn to the taxi way because fog had reduced his visibility.

Ripley changed her radio frequency back to Approach Control, and was given the following weather report. Indefinite ceiling at 200', sky obscured, visibility one and a half miles with fog.

Only a minute or two passes and the visibility further reduces to half a mile with fog. By now Ripley's flight 124 was back on the landing approach, three miles beyond the Outer Marker over Boston Harbor.

While she approached the landing runway a second time on instruments, three other commercial flights, a TWA and two Eastern Airline flights, request to divert to Bradley and Providence.

Boston Tower: "Okay, let me know when you have the approach lights in sight, please. Thank-you."
Flight 124: "I have 'em."
Boston Tower: "Corp Air 124, wind check zero seven zero at fourteen knots."
Flight 124: "One twenty-four."
Boston Tower: "Corp Air 124 the RVR (runway visual range [down the runway]) now twenty six hundred feet for four right."
Flight 124: "One twenty-four."
Boston Tower: "RVR twenty two hundred feet for Corp Air 124, four right."
No response.
Boston Tower: "Corp Air 124, tower."
No Response.
Boston Tower: "Corp Air 124, did ya get the lights, sir?"
No Response.
Intercom-2: (In house intercom) "Is the Corp Air down?"
Boston Tower: "Corp Air 124, say again."
No response.

CHAPTER TWENTY-TWO

Intercom-2: "Is Corp Air down?"
Boston Tower: "Corp Air 124, Boston Tower, how do you hear?"
No Response.
Boston Tower: "Corp Air 124, Boston Tower."
No Response.
Boston Tower: "No answer, Corp Air."
Port-25: "Boston Ground, Port two-five, holding short of 4 right and still no trace."
Boston Ground Control: "Port twenty-five, roger and, ah, d'ya happen to have, ah, any traffic in sight…landing on, ah, four right at all?"
Port-25: "Negative, none that we can see."

At 10:49 pm the Control Tower at Logan Airport lost contact with flight 124. The airport was closed, and search and rescue began immediately on the runways and grounds of Logan Airport. The fog, now on the ground, hindered efforts to find flight 124.

TWENTY-THREE

Photo by Sara or Bob Wolff.

The door to the closet squeaks as I open it wide and look inside. I find my favorite pair of pants, the green and gray plaid hip huggers, and the shirt that matches them, a navy blue body suit with the outline of a white whale stitched on the left side. I fold them both together, hang them over the orange painted closet door and swing the door nearly to a close. Janice is chatting away, as she prepares her things for the next day at school.

CHAPTER TWENTY-THREE

Tonight we are getting along well, and we both revel in the feeling of being sisters and being able to talk agreeably. I crawl under the covers of my twin bed and lay on my side facing Janice, who does the same on her bed, facing me. I feel disquiet.

"Do you ever wonder what will happen tomorrow?" I ask her.

"No," she replies, "I know exactly what is going to happen tomorrow. First, I'll probably oversleep, and rush around and then miss the bus anyway. Dad will give me a ride to school, and I'll stop at my locker...then..."

I drifted off to sleep, waking up when Dad steps into our room. I think, Oh no, I've overslept! and leap out of my warm bed, crossing the room to the closet door where my clothes are hung.

Dad sits down heavily on the bed saying he has something to tell us. I turn and everything that follows passes in slow motion. I look at him, his side and back to me as he faces my sister, under the covers in her twin bed . His hands are in front of him, his fingertips touching each other, his forearms resting on his knees as he leans over. Janice looks up toward him, her face sleepily showing from under the blankets, her long blond hair spread out on the pillow. He doesn't return her gaze, or look at me, but with his eyes to the floor he breathes, "Your mother had an accident last night."

How many seconds of silence pass for that sentence to be heard, processed, understood? Heard and processed, yes, understood, never. As though I have been in a car accident and knocked unconscious, the next several hours disappear from memory.

Last night after dinner Dad suggested that I go with Mom to work, but she said no. "It's a school night."

She called me into her room, and we stood at her bureau where so many times I had helped her curl her hair and watched her put on earrings. She opened her top drawer, and gave me a small box which contained a charm for my charm bracelet, engraved with my birth date from nine *days ago. She explained that it had taken her a while to find the right one, and then have it engraved. I think I hugged her. Then she left for work. Did she say goodbye?*

Reporters show up at the house and try to interview Dad while he is still absorbing the early morning news brought to him by the State Police. We receive word that a search in Boston Harbor has discovered wheel sections of the plane, and a one mile square area between Castle Island and the airport is being scoured by the Coast Guard, State Police and Airport personnel.

Not knowing where my mother is gives me the queerest feeling. Should I be sad she is gone or hopeful that she may be alive and waiting to be rescued? Emotionally I can't grasp the reason why she hasn't come home again. If I laugh at something funny, I feel guilty because my mother might be dead; if I cry I feel guilty because she might still be alive. Each hour that passes our fragile hopes drop a little lower, until we move about our tasks in a robot-like state of shock.

We hope and pray that she has somehow escaped and will be found on one of the nearby islands in Boston Harbor. But the rest of the plane, and its sole occupant, continue to elude the searchers.

Close friends and family arrive and take over the kitchen, meal preparation and picking up. Strangers at the door, the phone ringing nonstop, Mom's friends coming over and crying in their sorrow. Janice, Jeff and I feel underfoot; it isn't given to us kids the place to grieve, or the manner how. We hide in our rooms, unable to cope with the tremendous scene downstairs. We come down to get something to eat, to greet someone who has stopped by, or to see if there is any news, and then quickly retreat.

Amid boxes and a jumble of personal goods the day after moving, Jessica Lavey turned on the radio and heard that a small plane had gone down over Boston Harbor in the night. She turned the radio up and waited for more news, and within a short time

CHAPTER TWENTY-THREE

the report was updated to say it was a woman who had flown the plane.

Jessica hoped it wasn't Ripley, but the facts pointed to her. To comfort herself she thought about the last time they were together, and smiled recalling Ripley's statement that somewhere, someday, evidence would prove Bigfoot existed. But as the day wore on the news didn't improve, she began thinking, *Hold on Rip, hold on 'til they get to you,* and imagined her clinging to a buoy somewhere.

Jessica had moved several times since she and Ripley lived on the same street in Burlington. Now remarried to her husband, they had just moved from Boston to a new home in Plymouth. She and Rip had conferred about the proximity of the house to Plymouth Airport and how Ripley could fly there so the two could visit.

With the phone still not connected, her husband already at work with their only vehicle, Jessica waited anxiously that whole, long day. When he came home, she drove to the nearest pay phone to call and hear the awful news herself from Ken.

The next day she rented a car and drove to Wilmington. The house was full of people, everyone "just crushed with hurt," Jessica said. She visited as long as she could, then left with Dottie, driving to North Wilmington to the lot where Ripley and Kenneth had started building their dream house. It was partially framed, the woodwork standing up stark to the sky, open to the concrete poured slab floors -- as hollow as their feelings that day. Dottie walked to one end, and Jessica to the other.

They turned and found they had both been quietly crying. Dottie didn't have any tissues and Jessica had used all those she had.

Dottie said, "They'll have to do."

Jessica, horrified, responded with "No! You don't want to do that."

Dottie sharply said "Jessica, just give me the damned dirty tissues!"

Then they giggled a little, knowing the scene would have amused Ripley.

"Ripley was such a mother hen," Jessica recalled, "even though I was older than her. She taught me that it is better to throw up in the bathroom sink with water running rather than if you throw up in the toilet bowl, it just sits looking at you -- floating. She also taught me about thermal weather changes. She told me how much she loved to fly over fall foliage, saying, 'It looks like God mixed every color of paint and poured it over the forests.' "

TWENTY-FOUR

It's Tuesday, and the search and rescue teams have not found the plane. I overhear someone say, "It's no good having them hanging around here waiting. Send them to school where they can be busy." So we return to our lives at school on Wednesday. I try to act as though things at our house are normal, but everyone knows what has happened and what we are waiting for. True sympathy comes only from those we know best, or complete strangers.

The top story every night on television is coverage of the crash and the search. Newspapers carry the story on the front page, using pictures from past interviews, which appear like a sudden stab of pain to my chest as I shop for milk and bread with Dad at Weinberg's Corner Store. We stop to choose which paper we will buy, and a stranger overhears us, peers at me and says, "Is this your mother that is missing?"

I stow away my feelings of grief; Janice is angry, Jeff retreats to himself, my father appears lost. This week we fragment from a family into individuals, and we may never fully recover.

It is in math class when the Principal steps into the room. Everyone looks toward him, glad for any interruption from the lecture. He glances at me and an ocean starts to roar in my ears. I think he asks to dismiss me from class, but I cannot hear anything. Besides, I have already gathered my books and stood up to leave.

On Thursday the searchers employed the use of a sidesweep sonar, and that day the fuselage was located in 40 feet of water buried in silt. Ripley was still inside and covered under the debris of cargo, which had pitched forward on impact. The Coast Guard raised the plane, and the wreckage was brought to a hanger at Logan Airport and kept under guard until it could be inspected.

With the exception of "saltwater damage to the electronic components, precluding functional testing of the equipment,"[27] nothing else was found to be malfunctioning at the time of the crash. If the instruments had functioned incorrectly, we will never know.

An Investigator for the National Transportation Safety Board said he would not rule out the possibility of a malfunction in the runway ILS (Instrument Landing System) equipment, and that the airport records would have to be checked. The FAA closed Runway 4R and ordered the ILS to be re-certified.

In October, the Honorable Paul W. Cronin of the House of Representatives, requested information from the FAA regarding the Logan ILS and its safety of operation.

His reply came from Alexander P. Butterfield, Administrator with the FAA, who wrote:

CHAPTER TWENTY-FOUR

"The runway 4R ILS at Logan is similar to those in operation at over 250 airports throughout the country. Many are major hub airports. We have checked its performance record and find it to be at the same high level as the new-solid state ILS. Although it utilizes tubes, it is not considered antiquated and is maintained to our highest standards. We have no plans for replacing it or any other vacuum tube ILSs that provide a similar service at this time.

"Other aircraft utilizing the ILS 4R at Logan just prior to and soon after the accidents did not report any problem with the system. We have been unable to determine why either of the two flights descended below their minimums."

Earlier that year, July 31, a Delta DC9 had also crashed in Boston Harbor approaching the same runway in the same conditions, at a loss of 88 lives. The spokesman for Logan said there appeared to be no similarities between the two accidents.

EPILOGUE

I look over a worn, yellowed letter, the edges dry and cracked from age. Once again I am surprised at the depth of what I read.

"None of our lives will ever be the same again without Ripley. There is a definite kinship among all of us who fly, but there is something else we feel for certain people who inspire us and motivate us and it has nothing to do with what we do but what the person is. Ripley was such a person; our lives are richer for having known her."

Billie's words transport me back in time, to when I am a teenager opening the mail that arrives each day in a large bundle held together with an elastic.

The phone rings and I drop the mail on the white and gold speckled countertop. It's Millie, calling again to talk with Dad. The ladies of the Ninety-Nines have rallied together in the way they know best, by setting a goal to accomplish a worthy end and working steadily toward that goal. Jean Batchelder has spearheaded a memorial project and sent a letter to members.

Millie tells me of their ideas; a book for all the New England libraries or a simulator for the Museum of Science. They have established the Ripley Miller Memorial Fund, and Millie is the Treasurer. I listen interestedly, but immediately after I give up the phone to Dad my mind fills with thoughts of school, house chores, what to make for supper, and how did my favorite sweater end up in the dryer?

Thinking back I recall that within six months, the Ripley Miller Memorial Fund committee had outlined their goals and had begun reviewing ideas for the type of memorial they would put into place.

Virginia Bonesteel wrote of their goals: "First, to institute a program which will insure a permanent memorial to Ripley with some provision for an annual remembrance of Rip and others we have lost.

"Secondly, we want to promote women in aviation, and, in a greater sense, aviation in general.

"Lastly, we want to promote the Ninety-Nines. I believe that our project should transcend the Ninety-Nines. Aviation needs an educated general public.

"While we will personally remember Ripley the friend, fellow worker and pilot, we may best serve her memory by promoting aviation and the Ninety-Nines in a positive way."

One of the first steps they took was to immortalize her in an engraved granite plaque installed in the Forest of Friendship's Memory Lane. The International Forest of Friendship, "a living, growing memorial to the World History of Aviation and Aerospace," was a gift to America on her 200th birthday (1976) from the City of Atchison, Kansas, the birthplace of Amelia Earhart, and the Ninety-Nines.

> The Forest is nestled on a gentle slope, overlooking Lake Warnock, in the heartland of America. It is made up of trees representing all the fifty states and territories and thirty-five countries around the world...
> Among the special trees are one from George Washington's Mount Vernon Estate, the Bicentennial American Spruce, a tree from Earhart's grandfather's farm and the newly planted redbud from President Eisenhower's farm.
> Winding through the forest is Memory Lane, honoring those who have or still

are contributing to all facets of aviation and aerospace. Embedded in the concrete walk are granite plaques engraved with the names of more than 1200 honorees.
Included are such internationally recognized flyers as Amelia Earhart, Charles Lindbergh, Jeana Yeager, Rajiv Gandhi, the Wright Brothers, Sally Ride, Chuck Yeager, Beryl Markham, General "Jimmie" Doolittle, President Bush and Gen. Colin Powell.[28]

The biography of my mother sent to the Forest of Friendship by Mona Budding, was originally written by Virginia Bonesteel and appeared in the November 1973 issue of the Ninety-Nine News.

Virginia wrote: "Those of us who saw Ripley within recent weeks were struck by how relaxed and happy she was. Her enthusiasm at working for Corporate Air as a company pilot made it clear she had finally found a place in aviation that was challenging and rewarding for her. Ripley earned her success. I talked with Rip almost weekly during her training in the Baron in June. She expressed a compelling need to succeed, not just for herself, but for all of us. She knew that in a way she was representing women pilots and she didn't want to let us down."[29]

"Reading through our Ninety-Nines newsletter about the accomplishments of women in aviation makes me proud to be one of them,"[30] she had said. And so when *The History of the Ninety-Nines* was published in 1979, the committee had found the memorial they searched for.

With the funds collected since my mother's death, the committee purchased a dozen of these books. Among the twelve libraries that received books were Concord, NH, Augusta, ME, Montpelier, VT,

Hartford, CT, Boston, MA, the Community College of Rhode Island, Wakefield, MA and her home town of Wilmington.

We attended the simple ceremony held at the Boston Public Library and the Wilmington Memorial Library. Dad, Jeff, Janice, Billie and I stood on the dry lawn, squinting against the bright midday sun, and reminisced. Time was passing so quickly, and it appeared as though my mother was remembered only by us and a dwindling number of friends. I was wrong.

Many people had learned to fly through my mother's instruction, many more were introduced to flying from those students. The promotional articles and radio interviews had reached a few more. She shared a piece of her life -- her love of flying -- with everyone she met. Her memory remains in the lives of the people she came in contact with, who now share those memories with me.

My mother helped open doors, and they remained open. For other women, for younger women, for the next generation, for me. She did many things right, and a few things wrong in being all I needed her to be. My childhood memories, once separated into individual moments, now blend together with adult knowledge to capture an image of a remarkable woman to treasure.

Thank you, Mom.

ENDNOTES

[1] Reprinted with permission of *Woburn Daily Times*, from "Flying Beats Housework Any Day" 3/19/1965

[2] Reprinted with permission of *Woburn Daily Times*, from "Flying Beats Housework Any Day" 3/19/1965

[3] Ripley C. Miller

[4] Reprinted with permission of *Woburn Daily Times*, from "Flying Beats Housework Any Day" 3/19/1965

[5] Ninety-Nines *Thirty Sky Blue Years*, 30th Anniversary Program, from "A Glance Backward" by Kay Menges Brick, 1959

[6] Jaqueline Cochran, www.wasp-wwii.org/wasp/final/_report.htm; www.wasp-wwii.org/new/press/arnold_press1.htm *(12/7/1944)*, from the collection of *Marion C. Hanran, WASP,* 1/14/02

[7] From an AOPA (Aircraft Owners and Pilots Association) full page ad in the 1970 Ninety-Nines Convention program.

[8] Ninety-Nines Inc., *40th Anniversary Convention Program*, from "Welcome letter" by Nelson A. Rockefeller, 6/13/969

[9] *Ninety-Nine News*, from "Transcontinental Sky Trail," by Kay A. Brick, November-December 1969: Page 17

[10] Reprinted with permission of *The Lowell Sun*, from "Tewksbury flight instructor is a sandy-haired mother of three" by Ann Geib, 10/6/1969

[11] Reprinted with permission of *Wilmington Town Crier*, from "With Rip Miller It's Flying With A Difference" by Bob Morris 9/7/1972

[12] Reprinted with permission of *The Lowell Sun*, from "Let's take a plane...anywhere!" by Gerri Polli, 8/29/1971

[13] Reprinted with permission of *The Lowell Sun*, from "Tewksbury flight instructor is a sandy-haired mother of three" by Ann Geib, 10/6/1969

[14] Reprinted with permission of *The Lowell Sun*, from "Tewksbury flight instructor is a sandy-haired mother of three" by Ann Geib, 10/6/1969

[15] Reprinted with permission of *The Lowell Sun*, "Tewksbury flight instructor is a sandy-haired mother of three" by Ann Geib, 10/6/1969

[16] Reprinted with permission of *The Lowell Sun*, from "Let's take a plane...anywhere!" by Gerri Polli, 8/29/1971

[17] Reprinted with permission of *The Lowell Sun*, from "Tewksbury flight instructor is a sandy-haired mother of three" by Ann Geib, 10/6/1969

[18] Ninety-Nines Inc., *24th Annual Powder Puff Derby Official Program,* from "Thank Heaven" by Bob Buck, 1970

[19] ENE Chapter Ninety-Nines, Inc., *Wings in the Kitchen*, 1969

[20] ENE Chapter Ninety-Nines, Inc., *Wings in the Kitchen*, 1969

[21] Reprinted with permission of *The Lowell Sun*, from "Round Robin Looks for Precision" by Ann Geib, 5/3/1972

[22] Reprinted with permission of *The Woburn Daily Times*, from "Flying Beats Housework Any Day" 3/19/1965

[23] Reprinted with permission of *Boston Herald*, "Lady Airline Pilot Undaunted" by Barbara Rabinovitz, 9/3/1973

[24]Reprinted with permission of *Wilmington Town Crier*, from "With Rip Miller, It's Flying With a Difference" by Bob Morris, 9/7/1972
[25]Ripley C. Miller
[26]Reprinted with permission of *Boston Herald*, from "Lady Airlines Pilot Undaunted" by Barbara Rabinovitz, 9/3/1973
[27]National Transportation Safety Board, Aircraft Accident Report
[28]Reprinted with permission, Chairperson Fay Gillis Wells, Executive Director Kay Baker, from "The Forest and How It Came To Be" *International Forest of Friendship* (2002): Page 3
[29]*Ninety-Nine News*, November, 1973: Page 17
[30]*Boston Evening Globe*, from "Close-up" by Mary Sarah King, 7/8/1970

SOURCES

Thirty Sky Blue Years, 30th Anniversary Program, Ninety-Nines, Inc., Kay Brick, "A Glance Backward" 1959

The Ninety-Nines, Inc. 40th Anniversary Convention Program, New York-New Jersey Section Ninety-Nines, Inc.,1969

24th Annual Powder Puff Derby Program, Ninety-Nines, Inc., 1970

The Ninety-Nines Inc., 1970 International Convention Program, New England Section Ninety-Nines, Inc.,1970

Ninety-Nine News, publication by the Ninety-Nines Inc., Will Rogers World Airport, International Headquarters, Oklahoma City, OK 73159, 1969-1973

Welcome to AWNEAR, NE Section Ninety-Nines Inc., 5/1972

Wings in the Kitchen, ENE Chapter Ninety-Nines, Inc., 1969

The American Peoples Encyclopedia, Spencer Press Inc., Chicago, 1953

The Modern Family Health Guide, Edited by Morris Fishbein, MD, Doubleday & Company Inc., NY, 1959, pg. 78

The Fun of It, Amelia Earhart, The Junior Literary Guild and Brewer, Warren and Putnam, NY, 1932

A History of Women in America, Carol Hymowitz and Michaele Weissman, Bantam, NY, 1978

Boston Sunday Globe, "High Flying Gals Crowd Westfield Skies" Mary Sarah King, October 24, 1965

Woburn Daily Times, "Ripley Miller: Burlington's Only Aviatrix; Says Flying Beats Housework Any Day" March 19, 1965

Boston Herald American, "The Sky's Their Limit" Date unknown

Boston Evening Globe, "Housewives–Down in the Dumps? Take the Kids Up and Away!" Audrey Lynch, July 13, 1968

The Lowell Sun, "Tewksbury Flight Instructor is a Sandy-Haired Mother of Three" Ann Geib, October 6, 1969

Boston Evening Globe, "Close-Up*"* Mary Sarah King, July 8, 1970

The Lowell Sun (Sunday), *"*Let's Take a Plane–Anywhere!*"* Gerri Polli, August 29, 1971

The Lowell Sun, "Round Robin Looks for Precision" Ann Geib, May 3, 1972

Boston Sunday Globe, "22 Planes fly in All-Woman New England Air Race" Christina Robb, May 7, 1972

Boston Globe, "Women Pilots look to the Airlines" Jean Braucher, 1971

Wilmington Town Crier, "With Rip Miller–It's Flying With A Difference" Bob Morris, September 7, 1972

Boston Herald American, "Lady Airlines Pilot Undaunted" Barbara Rabinovitz, September 3, 1973

Wakefield Daily Item, March 13, 1937

AOPA Pilot, "Silver Screen Myths" Barry Schiff, February, 2003

National Aeronautic Association, http://www.naa-usa.org 7/2002

The Ninety-Nines Inc., http://www.ninety-nines.org 7/2002

National Aviation Hall of Fame, http://www.nationalaviation.org 7/2002

World War II WASP, http://www.wasp-wwii.org 7/2002

National Air and Space Museum, http://www.nasm.edu/nasm/aero/women_aviators 8/2002

www.aerofiles.com 7/2004

CNN, http://www.cnn.com/2002/us/northeast/10/06/acclaimed.women.ap/ 2/2003

Sky Prints, Aviation Enroute Atlas, John C. Mosby Jr. Pub, St. Louis, MO, 1978

Henry Ford Museum, www.hfmgv.org 9/2004

BoomTown, www.rextrailer.com 9/2004